Introduction

Chapter 1. Green Infrastructure and Urban Agriculture

Chapter 2. Rooftop Farms within Green Infrastructure

Chapter 3. Community Gardens within Green Infrastructure

Chapter 4. Case Studies

Chapter 5. Conclusion

Introduction

Green infrastructure refers to an interconnected network of natural and man-made elements within urban and suburban landscapes that work together to provide various ecological, economic, and social benefits. It's a multifaceted concept that goes beyond traditional infrastructure to integrate nature into the built environment.

At its core, green infrastructure encompasses everything from parks, gardens, and open spaces to green roofs, permeable pavements, and rain gardens. These elements contribute to environmental sustainability by enhancing biodiversity, improving air and water quality, reducing stormwater runoff, and moderating urban heat island effects.

Green infrastructure not only supports local ecosystems but also fosters human well-being. By providing accessible green spaces, it encourages outdoor activities, enhances mental health, and builds community connections. Economic benefits include increased property values, job creation, and reduced costs for stormwater management and climate change mitigation.

Moreover, green infrastructure is crucial in the face of climate change and urbanization. It offers adaptive strategies that enhance the resilience of cities, allowing them to cope with extreme weather events and changing environmental conditions. In contrast to traditional "gray" infrastructure, such as concrete roads and sewage systems, green infrastructure is often more flexible and adaptive, working with natural processes rather than against them.

The concept of green infrastructure has gained traction in urban planning and policy in recent years, reflecting a shift toward more sustainable and holistic approaches to city development. It represents a vital tool in creating livable, resilient, and vibrant urban areas, emphasizing a harmonious relationship between human society and the natural world. Whether through strategic landscaping, innovative building designs, or community-led garden projects, green infrastructure promotes a sustainable future, reimagining how cities can function and thrive.

Urban Agriculture as a Component of Green Infrastructure

Urban Agriculture, defined as the cultivation, processing, and distribution of food within urban areas, is an essential component of green infrastructure. Its significance in modern cities extends beyond mere food production, contributing to various ecological, social, and economic goals.

Ecologically, urban agriculture enhances green infrastructure by creating green spaces and increasing biodiversity within cities. Rooftop gardens, community plots, and vertical farms can replace impervious surfaces, improving water management and reducing heat island effects. These spaces can act as habitats for pollinators and other wildlife, promoting ecological balance.

Socially, urban agriculture fosters community engagement and empowerment. Community gardens, for example, can be a hub for social interaction, education, and collaboration. They allow city residents to connect with nature, each other, and the food they consume. This connection can lead to a greater sense of stewardship over local environments and an understanding of sustainable practices.

Economically, urban agriculture supports green infrastructure by providing local employment opportunities and reducing the need for long transportation chains for food products. By growing food locally, cities can decrease their carbon footprint, enhance food security, and stimulate local economies.

In the context of climate change and rapid urbanization, urban agriculture aligns with green infrastructure principles by offering adaptive and resilient solutions. It provides a multifunctional approach, addressing several urban challenges simultaneously. From enhancing the aesthetic appeal of urban areas to reducing reliance on external food sources, urban agriculture represents a dynamic and innovative aspect of green infrastructure. It's a manifestation of a shift towards a more integrated, sustainable urban planning model, where cities are not just places to live and work but thriving ecosystems that nourish both people and the planet.

Relevance in Sustainable City Development

The relevance of green infrastructure, including components such as urban agriculture, in sustainable city development is profound and multifaceted. As cities around the globe grapple with the challenges of urbanization, climate change, and resource constraints, the integration of green infrastructure provides solutions that are both resilient and environmentally responsible.

First and foremost, green infrastructure promotes ecological sustainability. By incorporating natural elements like parks, green roofs, and community gardens, cities can enhance biodiversity, improve air and water quality, and mitigate the urban heat island effect. These natural systems work in synergy with the built environment, reducing pollution and enhancing overall ecological balance.

Secondly, green infrastructure plays a vital role in social sustainability. Community gardens and accessible green spaces foster social interaction, promote community cohesion, and provide recreational opportunities. They also offer educational platforms to raise awareness about sustainable living and environmental stewardship.

Economically, green infrastructure can be a cost-effective approach to managing urban challenges. For instance, permeable pavements and rain gardens reduce the need for expensive stormwater management systems. Urban agriculture can contribute to local economies by creating jobs and stimulating small-scale local food production.

Furthermore, green infrastructure is essential in enhancing the resilience of cities. By embracing adaptive and flexible solutions that work with natural processes, cities are better equipped to withstand and respond to climate-related stresses and extreme weather events.

In essence, the relevance of green infrastructure in sustainable city development lies in its ability to align ecological, social, and economic objectives. It represents a paradigm shift from traditional urban planning towards a more integrated and holistic approach. By bridging the gap between urban life and nature, green infrastructure lays the groundwork for cities that are not only livable and vibrant but also sustainable and resilient in the face of future challenges. It is a key strategy in building cities that are in harmony with the natural environment and attuned to the well-being of their inhabitants.

Scope and Structure of the Book

The Scope and Structure of the book "Green Cities: Urban Agriculture, Rooftop Farms, and Community Gardens in Sustainable Infrastructure" provide a comprehensive and in-depth exploration of the integration of urban agriculture into the broader framework of green infrastructure.

Scope

The book's scope extends beyond mere definitions and theoretical concepts. It delves into practical applications and real-world examples, exploring how urban agriculture can be a catalyst for sustainable development in cities. Covering everything from design principles and economic models to legal considerations and community engagement, the book offers a multidimensional perspective on the subject. It reflects on the global trends, innovative practices, and provides an array of case studies demonstrating the successes and challenges in various cities around the world.

Structure

The book is organized into four main parts:

1. Green Infrastructure and Urban Agriculture: This section lays the foundational concepts, explains the relevance of urban agriculture in green infrastructure, and explores global trends and policies.
2. Rooftop Farms within Green Infrastructure: Focuses on the specifics of rooftop farming, including design, social aspects, maintenance, and future directions.
3. Community Gardens within Green Infrastructure: Examines the planning, development, community involvement, legal aspects, and long-term impact of community gardens as integral components of green infrastructure.
4. Case Studies and Models: This part highlights real-life examples and success stories, showcasing the practical implementation of the principles discussed throughout the book.

The conclusion sums up the insights gained and provides a visionary outlook on the future of urban agriculture within green infrastructure.

Through this well-defined scope and logical structure, the book aims to engage urban planners, policymakers, academics, community leaders, and general readers. It offers a compelling and actionable guide to harnessing the power of urban agriculture as a sustainable solution for modern cities, reinforcing the importance of green infrastructure in shaping resilient and vibrant urban landscapes.

Chapter 1. Green Infrastructure and Urban Agriculture

Green infrastructure and urban agriculture represent two vital concepts that are increasingly shaping the future of urban planning, design, and sustainability. In the heart of bustling cityscapes, these concepts provide a nexus for environmental stewardship, community well-being, economic vitality, and aesthetic enrichment. Chapter 1 embarks on a journey to explore the interconnectedness between green infrastructure and urban agriculture, unraveling their definitions, importance, applications, and synergies.

Green infrastructure encompasses a network of natural and semi-natural systems that provide essential services to urban areas, from stormwater management to climate mitigation and biodiversity conservation. Urban agriculture, on the other hand, is the cultivation, processing, and distribution of food within urban and peri-urban areas. Together, they create a mosaic of opportunities that transform urban spaces into thriving ecosystems that nourish both the land and its inhabitants.

This chapter lays the foundation for understanding the relationship between green infrastructure and urban agriculture, exploring how they complement and reinforce each other. Through examples, analyses, and insights, we will delve into the multifaceted benefits they offer to cities and their residents. The chapter also highlights the innovative practices, policy frameworks, and community engagements that are driving this transformation. Whether you are a city planner, a community activist, an urban farmer, or a curious citizen, this chapter offers a gateway into a world where green infrastructure and urban agriculture converge to create sustainable, resilient, and vibrant urban landscapes. Welcome to a journey of discovery, inspiration, and innovation.

Concepts and Principles

This section weaves together a tapestry of ideas that guide sustainable urban development, fostering a harmonious coexistence between nature, society, and the built environment.

What is Green Infrastructure?

Green infrastructure is a strategically planned and managed network of natural lands, working landscapes, and other open spaces that conserve ecosystem values and functions, while providing associated benefits to human populations. It's a concept that goes beyond traditional infrastructure, aiming to integrate nature into urban, suburban, and even rural areas, to create sustainable and resilient environments. Here's a detailed look at what green infrastructure entails:

1. Natural Elements: Green infrastructure incorporates natural elements such as parks, forests, rivers, and wetlands. These components support biodiversity by providing habitats for various species and enhancing ecological balance.
2. Built Features: It also includes built features like green roofs, permeable pavements, rain gardens, and urban farms. These innovative practices reduce stormwater runoff, filter pollutants, and can even produce food.
3. Connectivity: By creating connected networks of green spaces, green infrastructure facilitates the movement of wildlife and enhances the ecological connectivity across landscapes. This interconnectedness is essential for the resilience and adaptation of ecosystems.
4. Human Well-being: Green infrastructure promotes human well-being by providing recreational spaces, improving air quality, reducing noise pollution, and enhancing the aesthetic appeal of urban areas. Parks and green corridors encourage physical

activity, improve mental health, and foster community engagement.

5. Climate Mitigation and Adaptation: It plays a crucial role in mitigating climate change by sequestering carbon and reducing energy consumption through natural cooling. In terms of adaptation, green infrastructure enhances the ability of communities to withstand extreme weather events like floods and heatwaves.

6. Economic Benefits: There are significant economic benefits, such as increased property values near green spaces, job creation in landscaping and maintenance, and cost savings in stormwater management and climate change mitigation.

7. Sustainable Water Management: By capturing and filtering rainwater, green infrastructure supports sustainable water management, replenishing groundwater levels, and reducing the need for conventional stormwater infrastructure.

8. Community Integration: It fosters community integration by providing spaces for social interaction, community gardening, and environmental education.

9. Holistic Approach: Green infrastructure represents a holistic approach to urban planning, connecting different aspects of sustainability, resilience, health, and aesthetics.

10. Policy and Planning: Increasingly, municipalities and governments are integrating green infrastructure into urban planning and policy frameworks, recognizing its multifunctional benefits.

11. Challenges and Opportunities: While there are challenges such as maintenance, funding, and public awareness, the opportunities for innovation, collaboration, and long-term sustainability are immense.

In essence, green infrastructure is a dynamic and multifaceted concept that reimagines the way we view and interact with our surroundings. It's not just about adding greenery to urban areas but about creating synergies between natural and built environments. By aligning ecological integrity with human needs and urban functionality, green infrastructure is a key strategy for creating sustainable, resilient, and livable cities and communities. It represents a paradigm shift in planning and development, where nature is not an afterthought but an integral part of the design.

Role of Urban Agriculture

Urban agriculture plays an increasingly vital role in modern cities, addressing a variety of economic, social, and environmental challenges. Its significance extends far beyond mere food production, transforming urban landscapes and contributing to sustainability and resilience. Here's an in-depth look at the multifaceted role of urban agriculture:

1. Food Production: Urban agriculture facilitates local food production, providing fresh, nutritious, and often organic produce to urban dwellers. It reduces dependency on imported goods, enhancing food security.
2. Environmental Stewardship: Through practices like rooftop farming and community gardening, urban agriculture contributes to green infrastructure. It helps in carbon sequestration, waste recycling, and conserving biodiversity, promoting overall environmental health.
3. Community Engagement: Community gardens and collaborative farming projects foster social cohesion, enabling residents to work together, share knowledge, and create a sense of belonging and ownership in their neighborhoods.
4. Education and Awareness: Urban agriculture serves as an educational tool, teaching both children and adults about food origins, sustainable practices, and environmental stewardship. It can create a deeper connection with nature and food, even in densely populated areas.

5. Economic Opportunities: Local food production creates jobs and entrepreneurial opportunities. It can stimulate the local economy through farmers' markets, small-scale food processing, and related services.
6. Health and Well-being: Access to fresh produce promotes healthy eating habits, combating obesity and related health issues. The physical activity involved in gardening also contributes to overall well-being.
7. Climate Resilience: Urban agriculture contributes to climate resilience by reducing transportation-related emissions, managing stormwater through permeable surfaces, and providing shading and cooling effects in urban areas.
8. Land Use Efficiency: Utilizing underused spaces like rooftops, vacant lots, and balconies for farming maximizes land use efficiency, transforming otherwise neglected areas into productive green spaces.
9. Waste Reduction: Urban farming can involve composting and recycling of organic waste, reducing the burden on landfills and creating valuable soil amendments.
10. Policy Integration: More cities are recognizing the value of urban agriculture, integrating it into urban planning and zoning regulations. It's becoming an essential part of sustainable development strategies.
11. Challenges and Solutions: While urban agriculture faces challenges such as water scarcity, soil contamination, and limited space, innovative solutions and community-driven initiatives are continually emerging to address these issues.

In conclusion, the role of urban agriculture in today's cities is diverse and transformative. It's not just an alternative way of growing food but a multifunctional approach that connects people, enhances landscapes, supports local economies, and fosters sustainable living. Its integration into urban planning represents a paradigm shift towards more resilient, self-sufficient, and human-centered urban development. The potential impact of urban agriculture extends far beyond the garden plot, influencing how cities evolve, function, and thrive in an increasingly urbanized world.

Benefits and Challenges

Urban agriculture, as a crucial part of green infrastructure in cities, presents both notable benefits and inherent challenges. Here's a detailed examination of both aspects:

Benefits

1. Local Food Production: Urban agriculture ensures access to fresh, locally-sourced produce, reducing the reliance on distant food sources and transportation costs.
2. Environmental Sustainability: By transforming urban spaces into green areas, it mitigates climate change effects, promotes biodiversity, and helps in waste recycling.
3. Community Engagement: Community gardens and urban farms act as social hubs, enhancing community cohesion, and providing opportunities for education and recreation.
4. Economic Development: Urban agriculture creates jobs, stimulates local economies, and can increase property values in surrounding areas.
5. Health Improvement: Access to nutritious food and involvement in gardening activities can positively impact both physical and mental well-being.
6. Resilience: Urban agriculture contributes to the resilience of cities by providing local food sources during emergencies and adapting to environmental changes.
7. Land Utilization: By using vacant lots, rooftops, and other underutilized spaces, urban agriculture maximizes land use and beautifies urban landscapes.

Challenges

1. Space Constraints: Limited space in urban areas can restrict the scale of urban agriculture, demanding creative solutions like vertical farming.
2. Water Availability: Adequate water supply for irrigation can be a challenge, necessitating efficient water management and usage practices.
3. Soil Contamination: Urban soils may be contaminated with heavy metals or pollutants, requiring soil testing and careful selection of growing sites.
4. Regulatory Barriers: Zoning laws, building codes, and other regulations might restrict or complicate urban farming activities.
5. Financial Hurdles: The initial cost of setting up urban agricultural practices can be high, and ongoing maintenance can be resource-intensive.
6. Climate Factors: Specific urban microclimates may limit what can be grown, requiring careful planning and adaptation.
7. Social Acceptance: Gaining community acceptance and participation might be challenging, particularly in areas unfamiliar with urban agriculture practices.

In conclusion, the benefits and challenges of urban agriculture reflect its complex nature and its integration within the urban fabric. While it offers significant opportunities for sustainability, resilience, and community development, it also demands thoughtful planning, collaboration, and innovation to overcome inherent challenges. The interplay between these benefits and challenges shapes the effectiveness and success of urban agriculture as a transformative element in contemporary urban planning and development. It calls for a holistic approach that recognizes urban agriculture's multifaceted role in enhancing the quality of urban life. By understanding and addressing these challenges, cities can unlock the full potential of urban agriculture, harnessing its power to create more sustainable, healthy, and vibrant communities.

Global Trends and Policies

Green infrastructure and urban agriculture are shaping a new paradigm in urban planning, reflecting a worldwide shift towards sustainability, resilience, and community-centric development.

Green Infrastructure Initiatives

Green infrastructure Initiatives within the context of urban agriculture refer to the integration of farming and gardening practices into the urban environment to create sustainable, resilient, and vibrant cities. These initiatives blend agricultural activities with ecological design, emphasizing both productivity and environmental stewardship. Here's an overview of these initiatives:

1. Community Gardens and Urban Farms: Establishing community-based gardens and urban farms promotes local food production, enhances green spaces, fosters community engagement, and serves as an educational platform.
2. Green Roofs and Vertical Farming: Incorporating agriculture into building designs through green roofs and vertical farms transforms underutilized spaces, contributes to energy efficiency, and adds aesthetic value to urban areas.
3. Permaculture and Ecological Design: Integrating permaculture principles emphasizes a holistic approach to urban agriculture, focusing on water conservation, soil health, biodiversity, and sustainable practices.
4. Water Management: Developing rainwater harvesting, greywater recycling, and efficient irrigation systems in urban agriculture projects ensures responsible water usage and supports overall urban water management.
5. Waste Recycling: Implementing composting and anaerobic digestion in urban farming processes recycles organic waste,

reduces landfill burdens, and provides valuable soil amendments.

6. Urban Forestry and Edible Landscaping: Planting fruit trees and edible plants in public spaces beautifies urban areas and provides accessible food sources, enhancing both aesthetics and functionality.

7. Soil Remediation and Brownfield Redevelopment: Utilizing urban agriculture to remediate contaminated soils and rejuvenate brownfield sites transforms liabilities into community assets, promoting environmental restoration and sustainable land use.

8. Economic Development and Job Creation: Supporting urban agriculture through grants, incentives, and business development programs stimulates local economies, creates jobs, and encourages entrepreneurship.

9. Health and Well-being Promotion: Focusing on urban agriculture to enhance access to fresh produce and promote active lifestyles improves overall public health and well-being.

10. Policy Support and Integration: Formulating policies, zoning regulations, and incentives that specifically support urban agriculture ensures its successful integration into urban planning and development strategies.

11. Climate Change Mitigation and Adaptation: Incorporating urban agriculture within green infrastructure plans contributes to carbon sequestration, urban cooling, and climate resilience.

In conclusion, green infrastructure Initiatives in the context of urban agriculture represent a multifaceted approach to urban development that recognizes the interconnectedness of food production, environmental stewardship, social cohesion, and economic vitality. These initiatives highlight the transformative potential of urban agriculture as a tool for sustainable urban planning, weaving together natural systems and human activities in a harmonious and resilient urban fabric. By embracing urban agriculture within green infrastructure, cities can cultivate a future that is not only productive but also ecological, inclusive, and inspiring, redefining the relationship between urban dwellers and their environment.

Urban Agriculture Legislation

Urban agriculture Legislation refers to the laws, policies, and regulations that govern the practice of urban farming and gardening within city limits. As urban agriculture continues to gain prominence as a multifaceted solution to urban challenges, it necessitates clear and supportive legal frameworks. Here's an exploration of the key aspects of urban agriculture legislation:

1. Zoning Regulations: Zoning laws define where and how urban agriculture can be practiced within a city. They may outline specific areas for community gardens, rooftop farms, or aquaponic systems, balancing urban development with agricultural activities.
2. Land Use and Property Rights: Legislation must address land ownership, leasing, and access rights, ensuring that urban farmers have legal access to land and clarifying the rights and responsibilities of landowners and users.
3. Water Usage and Quality Control: Laws may regulate the sources, usage, and quality control of water in urban farming to protect public health and manage scarce water resources.
4. Soil Quality and Contamination Standards: Urban soils may be contaminated, so regulations must set standards for soil testing, remediation, and safe farming practices.
5. Building Codes and Safety Regulations: Building codes may govern the structural requirements for green roofs, vertical farms, and other infrastructure to ensure safety and compliance with local standards.

6. Sales and Distribution Laws: Regulations concerning the sale and distribution of locally grown produce may vary, affecting farmers' markets, community-supported agriculture, and direct-to-consumer sales.
7. Tax Incentives and Subsidies: Some cities provide tax incentives, grants, or subsidies to encourage urban agriculture, recognizing its social, economic, and environmental benefits.
8. Community Involvement and Participation: Legislation may foster community engagement by supporting community gardens, educational programs, and public participation in decision-making.
9. Integration with Other Policies: Urban agriculture legislation must align with other related policies, such as food safety, environmental protection, and economic development.
10. Challenges and Opportunities: Implementing and enforcing urban agriculture legislation may pose challenges, requiring clear definitions, inter-agency cooperation, and flexibility to adapt to local needs and innovations.
11. Global Perspectives: Different countries and cities have unique approaches to urban agriculture legislation, reflecting varying cultural, climatic, and urban contexts.

In conclusion, urban agriculture legislation serves as a vital framework to support, regulate, and promote urban farming. It plays a critical role in unlocking the potential of urban agriculture, providing legal clarity, and facilitating its integration into urban life. Effective legislation requires a nuanced understanding of the multifaceted nature of urban agriculture, ongoing dialogue between stakeholders, and a willingness to innovate and adapt to evolving urban landscapes. Urban agriculture legislation not only governs the practicalities of farming in cities but helps to shape a sustainable, resilient, and inclusive urban future, making it a central aspect of contemporary urban governance and planning.

Regional Differences

Regional differences play a significant role in shaping the practices, challenges, and opportunities of urban agriculture. Urban farming is influenced by various regional factors including climate, culture, economy, policy, and land availability. Here's an examination of these differences:

1. Climate and Geography: Different climates and geographies dictate what can be grown and when. Tropical regions may allow year-round cultivation, while colder areas may require specialized techniques like greenhouses or season extenders. Soil quality and topography also vary regionally, impacting what can be successfully cultivated.

2. Cultural Preferences and Traditions: Regional cultures influence food preferences, farming practices, and community involvement in urban agriculture. Traditional agricultural knowledge may guide urban farming in some areas, while others may be influenced by global food trends.

3. Economic Factors: Wealthier regions might emphasize urban agriculture as a tool for sustainability and community engagement, while lower-income areas may focus on food security and economic development. Available funding, market access, and consumer demand will also differ regionally.

4. Policy and Regulatory Environment: Regional governments may have varying levels of support and regulations related to urban agriculture. Some areas might have comprehensive urban farming policies, while others may have restrictive zoning laws or lack clear guidelines.

5. Technological Adoption and Innovation: The use of technology in urban agriculture, such as hydroponics or vertical farming, may vary based on regional investment in research, development, and education.

6. Land Availability and Use: Densely populated urban regions might emphasize rooftop farming or vertical gardens, while areas with more available land may focus on community gardens or larger urban farms.
7. Water Availability and Quality: Regions with abundant water may approach irrigation differently than arid regions, where water conservation practices are essential. Water quality regulations may also vary.
8. Social and Community Dynamics: Community engagement, education, and social organization in urban agriculture are influenced by regional social structures, values, and community needs.
9. Health and Safety Standards: Different regions may have varying health and safety regulations affecting urban agriculture, influencing practices related to soil contamination, pesticide use, and food handling.
10. Environmental Concerns and Sustainability Goals: Regional priorities related to environmental conservation, climate change mitigation, and sustainability will shape urban agriculture initiatives and practices.
11. Global Influences: International trade, global food markets, and multinational corporations may also shape regional urban agriculture practices.

In conclusion, regional differences in urban agriculture are multifaceted and deeply intertwined with the unique characteristics of each area. These differences necessitate a flexible and context-sensitive approach to planning, supporting, and implementing urban agriculture. Understanding and embracing these regional nuances allows for the development of urban agriculture systems that are tailored to local needs, opportunities, and challenges, contributing to more resilient, sustainable, and culturally relevant urban landscapes.

Synergies and Conflicts

Urban agriculture's integration with other green infrastructure elements is vital for building sustainable cities. This synergy creates multifaceted benefits, aligning agricultural practices with broader environmental, social, and economic goals. Here's how this integration can occur:

- Green Roofs and Vertical Gardens:
    - Incorporating gardens on rooftops and vertical surfaces.
    - Enhancing insulation, reducing energy consumption, and providing green space.
    - Managing stormwater runoff and improving air quality.
- Water Management Systems:
    - Utilizing rainwater harvesting and greywater recycling for irrigation.
    - Reducing pressure on municipal water supplies and promoting water conservation.
    - Connecting urban farming with wetland restoration and bioswale development.
- Waste Recycling and Composting:
    - Linking urban farming with organic waste collection and composting.
    - Transforming waste into valuable soil amendments.
    - Supporting circular economy principles and reducing landfill usage.
- Urban Forestry and Edible Landscaping:
    - Integrating fruit trees and edible plants in public spaces.
    - Enhancing urban biodiversity, aesthetics, and food accessibility.
    - Creating synergies with urban wildlife habitat restoration.
- Sustainable Transportation and Accessibility:

- o Connecting urban farms and gardens with pedestrian paths and bike lanes.
  - o Promoting local food distribution through farmers' markets near transit hubs.
  - o Enhancing community access to fresh, locally grown produce.
- Energy Efficiency and Renewable Energy:
  - o Designing urban agriculture sites with solar panels or wind turbines.
  - o Utilizing energy-efficient technologies in greenhouses or aquaponic systems.
  - o Aligning food production with broader urban sustainability goals.
- Community Engagement and Social Infrastructure:
  - o Collaborating with schools, community centers, and local organizations.
  - o Fostering education, volunteering, and social cohesion through urban farming.
  - o Building resilient communities with shared values and active participation.

In conclusion, integrating urban agriculture with other green infrastructure elements creates a cohesive approach to urban sustainability. It builds bridges between food production, environmental stewardship, community well-being, and economic development. This integration requires collaborative efforts, innovative thinking, and a commitment to holistic urban planning. By weaving urban agriculture into the fabric of green infrastructure, cities can cultivate a future that is not only green and productive but also resilient, inclusive, and harmonious, reflecting the interconnectedness of urban life and natural systems.

Potential Conflicts and Solutions

Urban agriculture's integration with other aspects of urban life and green infrastructure can sometimes lead to conflicts. Identifying these potential conflicts and proposing solutions is crucial for the successful implementation and sustainability of urban agriculture initiatives.

Potential Conflicts

- Land Use Conflicts:
  - Conflict: Competition for land between urban farming and other urban development needs like housing or commercial spaces.
  - Solution: Implement clear zoning regulations and promote multi-functional land use, such as combining urban farming with recreational spaces or parking areas.
- Water Usage and Quality:
  - Conflict: Excessive water usage in urban farming conflicting with water conservation goals.
  - Solution: Employ water-efficient practices such as drip irrigation, rainwater harvesting, and greywater recycling.
- Aesthetic Concerns:
  - Conflict: Perceived aesthetic issues with urban farming in certain locations, such as residential neighborhoods.
  - Solution: Establish design guidelines and encourage community participation in planning to create visually appealing urban farms.
- Health and Safety Concerns:
  - Conflict: Concerns over soil contamination, pesticide use, and food safety.
  - Solution: Implement and enforce strict health and safety regulations, provide training, and promote organic and sustainable farming practices.

- Economic Challenges:
  - o Conflict: Urban farming might be seen as economically unviable or in competition with traditional agriculture.
  - o Solution: Provide subsidies, tax incentives, and support for market access to ensure economic sustainability and complementarity with rural farming.
- Community Engagement:
  - o Conflict: Potential opposition or lack of engagement from community members.
  - o Solution: Facilitate community involvement in planning, decision-making, and maintenance to foster ownership and support.

- Environmental Impact:
  - o Conflict: Potential negative impacts on local ecosystems, such as monoculture practices.
  - o Solution: Promote biodiversity, ecological design, and environmentally responsible practices in urban agriculture.

Urban agriculture's integration with green infrastructure and urban development can lead to various conflicts. However, these conflicts can be managed and resolved through careful planning, clear regulations, community engagement, and a commitment to sustainability. By recognizing and addressing these potential conflicts proactively, urban agriculture can thrive as a harmonious and valuable part of urban life, contributing to a sustainable, resilient, and healthy urban environment.

Case Studies

Below are a few case studies that illustrate real-life examples of urban agriculture initiatives, highlighting various aspects, challenges, successes, and lessons learned.

New York City's Green Roof Initiative

- Location: New York City, USA.
- Overview: A push for green roof installations featuring urban gardens and farms.
- Successes: Reduction in building energy consumption, stormwater management, and enhancement of urban biodiversity.
- Challenges: Regulatory hurdles, high installation costs.
- Solutions: Implementation of tax abatements, grants, and policy support.

Havana's Urban Agriculture Movement:

- Location: Havana, Cuba.
- Overview: Widespread adoption of urban agriculture during the Special Period in the 1990s.
- Successes: Significant increase in local food production, community engagement.
- Challenges: Limited access to resources and technology.
- Solutions: Government support, community-driven initiatives, focus on organic practices.

Singapore's Sky Greens Vertical Farm

- Location: Singapore.
- Overview: A commercial vertical farm using innovative hydraulic-driven technology.
- Successes: High yield in a small footprint, water efficiency, local food supply.
- Challenges: High initial investment, technological complexity.
- Solutions: Government grants, technological innovation, market positioning.

Detroit's Urban Agriculture Ordinance

- Location: Detroit, USA.

- Overview: Comprehensive policy to support urban farming, community gardens, and green infrastructure.
- Successes: Legal clarity, support for community-led initiatives, revitalization of vacant lots.
- Challenges: Balancing urban development needs, land ownership issues.
- Solutions: Collaborative planning, clear zoning regulations, community engagement.

Melbourne's Community Rooftop Garden

- Location: Melbourne, Australia.
- Overview: Rooftop community garden promoting local food, social interaction, and sustainability.
- Successes: Community engagement, education, aesthetics.
- Challenges: Water access, building regulations.
- Solutions: Rainwater harvesting, collaboration with building owners, community-driven management.

Nairobi's Urban Farming for Food Security:

- Location: Nairobi, Kenya.
- Overview: Promotion of small-scale urban farming to enhance food security in low-income areas.
- Successes: Increased access to fresh produce, empowerment of women farmers.
- Challenges: Land tenure issues, water scarcity.
- Solutions: Legal support, community organization, water-efficient practices.

These case studies demonstrate the diversity and potential of urban agriculture across different contexts. They highlight the importance of a holistic approach, integrating community engagement, policy support, innovation, and sustainability. By learning from these real-life examples, cities and communities can develop tailored urban agriculture initiatives that respond to local needs and opportunities, contributing to urban resilience and sustainability.

Chapter 2. Rooftop Farms within Green Infrastructure

Chapter 2 focuses on one of the most innovative and captivating aspects of urban agriculture: rooftop farms within green infrastructure. As cities expand and open spaces become more scarce, rooftops have emerged as a frontier for agricultural development, transforming underutilized areas into thriving gardens and food production centers.

Rooftop farms are more than just an aesthetic addition to the urban landscape; they represent a confluence of sustainability, community engagement, economic viability, and architectural creativity. This chapter explores the multifaceted dimensions of rooftop farming, from structural considerations and water management to plant selection, biodiversity, and community involvement.

We will take a closer look at the benefits that rooftop farms offer to the urban environment, such as mitigating heat island effects, enhancing biodiversity, reducing stormwater runoff, and providing local food sources. We'll also explore the challenges and solutions associated with implementing rooftop farms, such as navigating regulations, ensuring structural integrity, and creating economically sustainable models.

The integration of rooftop farms within the broader context of green infrastructure exemplifies a harmonious blend of functionality and beauty, practicality and vision. This chapter provides insights, examples, and guidelines for those interested in exploring rooftop farming as part of a comprehensive approach to urban sustainability.

Introduction to Rooftop Farming

Rooftop farms represent an innovative approach to urban agriculture that directly aligns with the principles of green infrastructure. They play a crucial role in transforming unused rooftop spaces into productive landscapes, contributing to the sustainable development of urban areas. Here's how rooftop farms connect to the broader framework of green infrastructure:

- Space Utilization:
    - Efficient Land Use: In densely populated urban areas, space is a valuable commodity. Rooftop farms utilize otherwise wasted space, offering a solution to the land scarcity issue.
    - Multi-Functional Spaces: Many rooftop farms are designed to be multi-purpose, serving as community gathering spaces, educational hubs, or recreational areas.
- Environmental Benefits:
    - Climate Regulation: Rooftop farms can contribute to climate regulation by providing insulation, reducing building energy consumption for heating and cooling.
    - Stormwater Management: They act as a natural sponge, absorbing rainwater, thus helping in controlling stormwater runoff and reducing the strain on urban drainage systems.
    - Biodiversity Enhancement: By creating habitats for pollinators and other species, rooftop farms increase urban biodiversity.
- Social and Community Impact:
    - Community Engagement: Rooftop farms often engage the community through volunteer opportunities,

educational programs, and social events, fostering a sense of ownership and connection.
  - o Local Food Production: They provide access to fresh, locally grown produce, enhancing urban food security and reducing the need for transportation, thereby cutting carbon emissions.
- Economic Considerations:
  - o Job Creation: Rooftop farming initiatives can create jobs and business opportunities within urban areas.
  - o Property Value Enhancement: Well-designed rooftop gardens can increase property values by improving aesthetics and providing additional functional space.

- Integration with Other Green Technologies:
  - o Synergy with Renewable Energy: Some rooftop farms integrate solar panels or wind turbines, harmonizing food production with renewable energy generation.
  - o Water Efficiency: Many incorporate rainwater harvesting or greywater recycling systems, contributing to water conservation goals.
- Policy and Planning Alignment: Cities like Toronto, New York, and Singapore have recognized the value of rooftop farms and incorporated them into urban planning and policies, offering incentives or regulations to support their development.
- Challenges and Solutions:
  - o Structural Considerations: Ensuring that buildings can support the weight and other requirements of a rooftop farm is a significant challenge, necessitating proper assessment and engineering solutions.
  - o Access and Regulation: Navigating building codes, regulations, and ensuring accessibility may require careful planning and collaboration with local authorities.

In conclusion, rooftop farms represent a convergence point for many aspects of green infrastructure, reflecting a holistic approach to urban sustainability. They embody a vision where urban areas are not just concrete jungles but thriving ecosystems that nourish the body, mind, and soul of their inhabitants. As cities continue to grow and face unprecedented challenges, the integration of rooftop farms into the fabric of urban life may prove to be not just an attractive option but a necessary strategy for a sustainable future.

Benefits to Urban Environment

Rooftop farms are gaining popularity in urban environments, transforming barren rooftops into lush, productive spaces. Their introduction to cityscapes provides various benefits that extend beyond food production and align with the broader goals of sustainable urban living.

- Environmental Impact:
    - Energy Efficiency: Rooftop farms act as natural insulators, helping to regulate building temperature and reduce energy consumption for heating and cooling.
    - Air Quality Improvement: The plants in rooftop farms filter air pollutants, contributing to cleaner urban air.
    - Stormwater Management: These farms absorb rainwater, reducing stormwater runoff and relieving stress on sewage systems.
- Social Benefits:
    - Community Engagement: They provide community gathering spaces, fostering social connections, and cohesion. Local residents can participate in growing food, attending workshops, or enjoying recreational activities.

- o Educational Opportunities: Rooftop farms serve as living laboratories, offering hands-on learning experiences about agriculture, sustainability, and ecology.
- Health and Well-being:
  - o Access to Fresh Produce: By providing locally sourced, fresh produce, rooftop farms enhance urban food security and promote healthier eating habits.
  - o Mental Health Benefits: Green spaces, including rooftop farms, are associated with improved mental well-being, providing respite from urban stress and opportunities for relaxation and mindfulness.
- Economic Advantages:
  - o Job Creation: Rooftop farming creates opportunities for employment, entrepreneurship, and economic development within urban areas.
  - o Increase in Property Value: The aesthetic appeal and functionality of rooftop farms can add value to properties.
- Biodiversity Enhancement:
  - o Habitats for Wildlife: Rooftop farms can serve as habitats for birds, insects, and other urban wildlife, supporting biodiversity within the city.
  - o Promotion of Pollinators: The presence of flowering plants can attract pollinators, vital for many ecosystems.

- Integration and Sustainability:
  - o Part of a Circular Economy: Many rooftop farms practice composting and rainwater harvesting, reflecting principles of resource efficiency and circular economy.
  - o Alignment with Sustainability Goals: By contributing to environmental protection, social well-being, and economic growth, rooftop farms support broader urban sustainability goals and strategies.

- Challenges and Opportunities:
  - o Navigating Regulations and Standards: While some cities actively support rooftop farming, others may have regulatory barriers that require careful navigation.
  - o Innovation and Design: The need for suitable structural design and farming techniques presents challenges but also opportunities for innovation and collaboration between architects, engineers, farmers, and urban planners.

In summary, rooftop farms present a multifaceted solution to some of the most pressing urban challenges. They offer a vision of cities where buildings not only house people and businesses but also contribute to a vibrant, sustainable, and resilient urban ecosystem. By embracing rooftop farms, cities can take a significant step toward reconciling urban living with nature, fostering communities that thrive not in spite of their urban surroundings but in harmony with them.

Case Studies

Rooftop farms have been implemented globally, examples of which include the following.

Brooklyn Grange, New York City, USA

- Overview: One of the largest rooftop soil farms in the world, spanning over 2.5 acres across two rooftops.
- Benefits: Produces over 50,000 pounds of organically cultivated produce annually; hosts educational programs; and has a significant stormwater retention capacity.
- Challenges: Navigating building regulations and securing initial funding.
- Solutions: Collaboration with building owners, securing grants, and community support.

Lufa Farms, Montreal, Canada

- Overview: A pioneering rooftop greenhouse project, utilizing hydroponic systems to grow a variety of crops.
- Benefits: Year-round production, reduction in transportation emissions through local distribution, and efficient water usage.
- Challenges: Climate control and energy efficiency.
- Solutions: Technological innovation, including the integration of renewable energy sources.

Sky Greens, Singapore

- Overview: Vertical farming system utilizing hydraulic-driven rotating towers.
- Benefits: High yield within a small footprint, reduced water and fertilizer usage, and contribution to local food security.
- Challenges: High initial investment and technological complexity.
- Solutions: Government support, technological advancements, and targeted market positioning.

The Food Roof Farm, St. Louis, USA

- Overview: A community-driven rooftop farm focusing on community engagement and environmental education.
- Benefits: Community involvement, youth education, stormwater management, and support for food banks.
- Challenges: Community outreach and volunteer retention.
- Solutions: Regular events, workshops, and strong community partnerships.

Sundrop Farms, Adelaide, Australia

- Overview: A pioneering greenhouse facility situated on a rooftop, utilizing solar power and seawater to produce crops.

- Benefits: Unique desert location uses solar energy to desalinate seawater for irrigation, achieving a sustainable water source; produces tomatoes for over 100,000 people annually.
- Challenges: Harsh environmental conditions and the complexity of integrating multiple sustainable technologies.
- Solutions: Innovative design and engineering, collaboration with researchers, and targeted government support.

These case studies represent a snapshot of the varied and innovative approaches to rooftop farming around the world. From community engagement to technological advancements, these initiatives showcase the potential of rooftop farms to transform urban landscapes and contribute to sustainable development. The lessons learned from these examples can inspire and guide future endeavors in cities seeking to integrate agriculture into the fabric of urban life, thus enhancing resilience, sustainability, and community well-being.

Design and Implementation

In the context of rooftop farms, design and implementation serve as the crucial blueprint and action plan, harmoniously weaving together structural engineering, ecological principles, aesthetic considerations, and community needs to transform urban rooftops into thriving agricultural landscapes.

Structural Considerations

Structural considerations are paramount when planning and implementing rooftop farms, as they must be designed to ensure safety, efficiency, and functionality. Here's an exploration of these aspects:

- Weight Bearing Capacity:
    - Assessment: Before a rooftop farm can be installed, a structural assessment must be conducted to determine if

the building can support the additional weight of soil, plants, water, equipment, and people.

- o Solutions: Depending on the building's capacity, lightweight growing mediums and careful selection of plants can help align the farm with the building's structural limits.
- Waterproofing and Drainage:
  - o Water Management: Rooftop farms require proper waterproofing and drainage systems to prevent water leakage, which can cause structural damage.
  - o Solutions: High-quality waterproof membranes and carefully designed drainage paths can prevent potential issues.
- Wind and Climate Considerations:
  - o Wind Resistance: Rooftops are often exposed to higher wind speeds, and the design must consider the potential impact on plants and structures.
  - o Climate Adaptation: Depending on the location, the design might also need to account for snow loads, extreme temperatures, or other climatic conditions.
  - o Solutions: Windbreaks, appropriate plant selection, and seasonally adaptive designs can mitigate these challenges.
- Accessibility and Safety:
  - o Safe Access: Ensuring safe access for workers, visitors, and emergency services is a crucial aspect of design.
  - o Railings and Walkways: Safety measures like railings and proper walkways must be integrated.
  - o Solutions: Collaboration with safety experts and adherence to local building codes can ensure a safe environment.
- Integration with Building Systems:

- Heating and Cooling: The rooftop farm may affect the building's heating and cooling systems and vice versa.
- Utility Connections: Provisioning for water, electricity, and other utilities must be done thoughtfully.
- Solutions: A holistic design approach, considering the entire building's function and performance, can create synergies rather than conflicts.
- Aesthetic and Functional Design:
  - Visual Appeal: Rooftop farms should be visually appealing and align with the building's overall aesthetics.
  - Multi-Functional Spaces: They can be designed to serve multiple purposes, such as recreation or community gathering.
  - Solutions: Involving architects, landscape designers, and the community in the planning process can achieve a balanced and pleasing design.

Structural considerations for rooftop farms are complex and multifaceted, requiring careful planning, collaboration among various experts, and adherence to building codes and regulations. While these considerations present challenges, they are not insurmountable. Innovative designs, appropriate technology, and attention to detail can lead to rooftop farms that are not only productive and sustainable but also safe, beautiful, and integrated with the urban environment. Such endeavors can transform rooftops into vibrant spaces that contribute to the broader goals of urban sustainability and resilience.

Water Management and Sustainability

Water management is a critical aspect of rooftop farming, directly affecting both sustainability and productivity. Efficient water management can significantly reduce consumption, minimize waste, and promote the overall health of the urban ecosystem. Here are the primary considerations:

- Water Sources and Efficiency:
  - Rainwater Harvesting: Collecting and storing rainwater is a common and sustainable practice for rooftop farms. Rainwater can be channeled into storage tanks and used for irrigation.
  - Recycled Water: Utilizing treated greywater can also be an effective strategy.
  - Irrigation Systems: Drip or other efficient irrigation systems can minimize water usage, delivering water directly to the plant roots.
  - Solutions: Combining these practices can create a closed-loop system that maximizes efficiency and minimizes waste.
- Water Quality:
  - Testing and Treatment: Regular monitoring and potential treatment of collected rainwater or recycled water is necessary to ensure it meets agricultural standards.
  - Solutions: Implementing filtration and purification techniques can ensure water quality for plant health.
- Drainage and Runoff Management:
  - Preventing Leakage: Proper waterproofing and drainage planning prevent water leakage into the building and excessive runoff.
  - Runoff Treatment: Consideration for treating or utilizing the nutrient-rich runoff can enhance sustainability.
  - Solutions: Integrating drainage systems with water collection and treatment can create a holistic water management cycle.

- Climate Considerations:
  - o Climate Adaptation: Water management strategies must consider local climate factors, such as rainfall patterns and seasonal variations.
  - o Solutions: Tailoring water management practices to local conditions can enhance resilience and sustainability.

- Ecological Impact:
  - o Local Water Ecosystems: Rooftop farms' water management practices should be designed to minimize negative impacts on local water bodies and ecosystems.
  - o Solutions: Strategic planning and adherence to environmental regulations can mitigate potential risks.
- Integration with Other Systems:
  - o Building Integration: Coordination with building's existing water systems is necessary to ensure compatibility.
  - o Solutions: Collaborative planning between architects, engineers, and urban farmers can achieve seamless integration.

Water management in rooftop farming is a nuanced and vital aspect, closely tied to sustainability goals. By utilizing rainwater harvesting, efficient irrigation, and thoughtful design, rooftop farms can model responsible water stewardship. Such practices not only enhance the productivity of the farms but also contribute to broader urban sustainability objectives, demonstrating how agriculture and ecology can be integrated into the urban fabric. These approaches exemplify how innovation and thoughtful planning can transform a basic necessity like water into an opportunity for positive environmental impact, connecting urban residents to natural cycles, and fostering a more resilient and sustainable urban environment.

Plant Selection and Biodiversity

Plant selection and biodiversity play essential roles in rooftop farming. They can significantly influence productivity, sustainability, and the overall ecosystem of a rooftop garden. Here are key considerations and strategies for making informed choices:

- Aligning with Goals:
    - o Food Production: For farms focused on food production, selecting high-yielding, disease-resistant, and locally adapted crops is critical.
    - o Aesthetic or Recreational Spaces: If the rooftop farm serves a decorative or community purpose, ornamental and native plants might be prioritized.
    - o Solutions: Clearly defining the farm's primary objectives will guide the initial plant selection process.
- Climate and Microclimate Considerations:
    - o Weather Conditions: Understanding local climate and the specific microclimate of the rooftop (e.g., wind, sun exposure) informs plant selection.
    - o Seasonal Changes: Considering seasonal variations and selecting plants accordingly can ensure year-round success.
    - o Solutions: Conducting a thorough climate analysis and selecting appropriate plant varieties can optimize growth and resilience.
- Soil and Growing Medium:
    - o Soil Quality: Plants must be matched to the soil or growing medium used, considering factors like pH, nutrient content, and weight constraints of the building.
    - o Solutions: Regular soil testing and amendments can align the soil's characteristics with plant needs.
- Biodiversity for Ecological Balance:

- o Pest Control: Cultivating a variety of plants can attract beneficial insects and create a natural defense against pests.
- o Soil Health: Diverse plantings can enhance soil health and nutrient cycling.
- o Solutions: Implementing polyculture or companion planting techniques can foster a balanced and resilient ecosystem.
- Community Engagement and Cultural Relevance:
  - o Cultural Preferences: In community-driven projects, selecting plants that reflect local culture and culinary preferences can enhance community engagement.
  - o Solutions: Engaging community members in plant selection can foster ownership and success.
- Environmental and Sustainability Considerations:
  - o Native Plants: Utilizing native plants can support local biodiversity and require less maintenance.
  - o Water Efficiency: Selecting drought-tolerant species can enhance water conservation efforts.
  - o Solutions: Aligning plant selection with broader sustainability goals can reinforce the farm's positive environmental impact.

Plant selection and biodiversity in rooftop farming are complex and multifaceted considerations that require a thoughtful and strategic approach. Aligning plant choices with the farm's goals, local climate, soil conditions, community preferences, and sustainability objectives ensures that the rooftop farm thrives and contributes positively to the urban environment. By embracing diversity and leveraging local knowledge, rooftop farms can become not just sources of food, but vital hubs of ecological stewardship, community connection, and cultural expression, weaving nature's bounty into the fabric of urban life.

Social and Economic Aspects

The social and economic Aspects of rooftop farming extend beyond the soil and seeds, encompassing community engagement, job creation, local economy stimulation, and the enhancement of social well-being within urban areas.

Community Engagement

Community engagement in rooftop farming is not only a means of fostering social connections but also a vital component in the success and sustainability of these urban agricultural projects. Here's an exploration of how community engagement is integrated into rooftop farming:

- Creating Community Spaces:
  - Inclusivity: Rooftop farms can serve as shared spaces where community members gather, interact, and collaborate.
  - Education and Workshops: Offering workshops, tours, and educational programs can empower the community with knowledge about agriculture, nutrition, and sustainability.
  - Solutions: Hosting regular community events and educational sessions can create an inclusive and vibrant community hub.
- Local Employment and Volunteering Opportunities:
  - Skill Development: Rooftop farms can provide opportunities for skill development, training, and employment in urban agriculture.
  - Volunteer Programs: Engaging volunteers can foster a sense of ownership and strengthen community bonds.

- o Solutions: Creating structured volunteer programs and providing training can enhance participation and skill-building.
- Collaboration with Local Organizations:
  - o Partnerships: Collaborating with schools, NGOs, local businesses, and government agencies can amplify the impact of rooftop farming.
  - o Solutions: Establishing clear partnership agreements and shared goals can facilitate successful collaborations.
- Community-Driven Decision Making:
  - o Participation in Planning: Involving community members in planning, design, and decision-making processes ensures that the farm aligns with local needs and preferences.
  - o Solutions: Regular community meetings and transparent communication channels can foster active participation.

- Promoting Health and Well-being:
  - o Access to Fresh Produce: Rooftop farms can provide local access to fresh, healthy produce, addressing food deserts in urban areas.
  - o Therapeutic Benefits: The act of gardening itself can have therapeutic benefits, enhancing mental well-being.
  - o Solutions: Integrating with local health initiatives and providing affordable access to produce can enhance community well-being.
- Cultural Connections and Social Integration:
  - o Cultural Expression: Rooftop farms can reflect the cultural diversity of the community through plant selection, design, and culinary programs.
  - o Social Integration: They can serve as neutral spaces where people from various backgrounds come together, fostering social cohesion.

o   Solutions: Encouraging cultural programs and inclusive design can strengthen social integration.

Community engagement in rooftop farming is multifaceted and transformative, turning urban spaces into thriving social and ecological hubs. By fostering collaboration, inclusivity, education, and empowerment, rooftop farms can become more than mere agricultural endeavors. They can be centers of innovation, community development, and cultural expression, weaving together the threads of urban life into a vibrant tapestry of sustainability and connection. It underscores the notion that food production is not just a solitary pursuit but a shared endeavor that nourishes the body, mind, and community alike.

Economic Models

Rooftop farming presents various economic opportunities and challenges that require careful planning and innovative approaches. The following exploration delves into different economic models that can be employed in rooftop farming, highlighting key considerations:

- Commercial Farming Model:
    o   For-Profit Approach: Some rooftop farms are designed with a profit-making motive, selling produce to local markets, restaurants, or directly to consumers.
    o   Scaling and Distribution: This model may require considerable investment in infrastructure, scaling, and distribution channels.
    o   Solutions: Collaborating with local businesses and implementing efficient production techniques can enhance profitability.
- Community-Supported Agriculture (CSA) Model:
    o   Subscription-Based Farming: Community members can subscribe to receive a regular share of the produce, thus providing consistent revenue for the farm.

- o Community Engagement: CSAs often foster a stronger connection between farmers and consumers.
  - o Solutions: Transparent communication and flexibility in subscription plans can enhance community participation.
- Social Enterprise Model:
  - o Dual Mission: This model combines financial goals with social or environmental objectives, such as job training, community engagement, or environmental stewardship.
  - o Funding Sources: Grants, donations, and impact investing may support this model.
  - o Solutions: Developing a clear mission and metrics for social impact can attract support from various stakeholders.
- Educational and Non-Profit Model:
  - o Focus on Education and Outreach: Some rooftop farms operate as non-profit entities focusing on education, research, and community engagement rather than profit.
  - o Funding Challenges: Reliance on grants, donations, and volunteers can create financial instability.
  - o Solutions: Building strong partnerships and diversifying income streams, such as hosting events or workshops, can enhance sustainability.

- Hybrid Models:
  - o Combining Approaches: Many rooftop farms combine various economic models to create a balanced and resilient economic structure.
  - o Solutions: Flexibility and adaptability in financial planning can lead to a more sustainable economic model.
- Challenges and Considerations:

- High Initial Costs: The start-up costs for rooftop farming, including infrastructure, soil, and water systems, can be substantial.
- Regulatory Compliance: Navigating zoning laws, building codes, and other regulations can add complexity.
- Solutions: Thorough planning, collaboration with experts, and engagement with local authorities can mitigate challenges.

Rooftop farming's economic models are as diverse and multifaceted as the farms themselves. By aligning with clear goals, whether commercial, community-driven, or educational, and by thoughtfully navigating challenges, rooftop farms can thrive as vital parts of the urban economic landscape. Creativity, adaptability, and a keen understanding of local needs and resources are key ingredients in crafting an economic model that nourishes not only the farm but the broader community and urban ecosystem. Whether forging a path to profit, education, or social impact, rooftop farms offer fertile ground for innovation and growth, both economically and sustainably.

Policy Framework

The development and success of urban agriculture, particularly rooftop farming, are significantly influenced by the policy framework. The right policies can facilitate growth, innovation, and sustainability, while the lack of supportive legislation can hinder progress. Here's an exploration of the policy framework relevant to rooftop farming:

- Zoning Regulations and Building Codes:
  - Inclusion in Zoning Laws: Specific zoning laws allowing or encouraging rooftop farming can facilitate its development.

- o Structural Considerations: Building codes must accommodate the unique structural needs of rooftop farms, considering weight, safety, and accessibility.
- o Solutions: Collaborating with urban planners and local authorities to develop supportive zoning and building codes.
- Incentives and Subsidies:
  - o Financial Support: Grants, tax incentives, or subsidies can lower the financial barriers to starting a rooftop farm.
  - o Sustainability Incentives: Incentives for environmentally sustainable practices, like rainwater harvesting or using renewable energy, can enhance sustainability.
  - o Solutions: Creating clear guidelines for eligibility and an accessible application process for incentives.
- Water and Waste Regulations:
  - o Water Usage Policies: Guidelines for using recycled or harvested rainwater can support sustainable water management in rooftop farming.
  - o Composting and Waste Handling: Regulations supporting on-site composting and waste reduction can enhance sustainability.
  - o Solutions: Developing comprehensive guidelines that align with broader urban sustainability goals.
- Health and Safety Regulations:
  - o Food Safety Standards: Ensuring that rooftop farms comply with food safety regulations is vital for protecting public health.
  - o Occupational Safety: Policies must also consider the safety of those working on or visiting the rooftop farms.
  - o Solutions: Regular inspections and guidance on best practices can maintain high safety standards.

- Integration with Urban Planning:
    - o Comprehensive Urban Agriculture Policies: Integrating rooftop farming within broader urban planning can create synergies with other sustainability and community development goals.
    - o Solutions: Collaboration between various city departments and stakeholders in policy development.
- Community Engagement and Equity:
    - o Access and Inclusion: Policies must ensure that the benefits of rooftop farming, such as access to fresh produce or green spaces, are equitably distributed.
    - o Solutions: Engaging communities in policy formulation and decision-making processes.

The policy framework for rooftop farming requires a multidimensional approach that recognizes its unique characteristics and potential contributions to urban life. Crafting supportive policies requires collaboration between government agencies, urban farmers, community organizations, and other stakeholders. From zoning laws to sustainability incentives, and from safety regulations to community engagement, a coherent policy framework can turn rooftop farming from a niche practice into a mainstream feature of sustainable urban development. The challenge lies in aligning the multitude of interests and considerations, but the reward is a thriving urban landscape where agriculture and city life grow in harmony.

Maintenance and Sustainability

Maintenance and sustainability in rooftop farming are fundamental to ensuring long-term success, requiring careful planning, ongoing care, and a commitment to ecological principles that support both the farm and the broader urban environment.

Ongoing Care

Ongoing care is a fundamental aspect of rooftop farming that ensures the long-term sustainability and productivity of the farm. It involves regular monitoring, maintenance, and nurturing of both the physical infrastructure and the plants. Here's an exploration of key aspects of ongoing care in rooftop farming:

- Soil Management and Fertility:
  - Soil Testing: Regular testing for nutrient levels and pH ensures that the soil maintains its fertility.
  - Composting: Incorporating organic matter like compost helps maintain soil structure and nutrient content.
  - Solutions: Implementing crop rotation and utilizing organic fertilizers can support soil health.
- Water Management:
  - Irrigation Systems: Regular maintenance of irrigation systems ensures efficient water usage.
  - Rainwater Harvesting: Rooftop farms often use harvested rainwater, requiring ongoing care of collection and filtration systems.
  - Solutions: Investing in water-saving technologies and regular monitoring can prevent over- or under-watering.
- Pest and Disease Control:
  - Integrated Pest Management: Ongoing monitoring for pests and diseases allows for timely intervention using environmentally friendly methods.
  - Solutions: Encouraging beneficial insects and using organic pesticides can minimize harmful impacts.
- Structural Integrity and Safety:
  - Load-Bearing Checks: Regular inspections of structural elements ensure the rooftop can sustain the load.
  - Accessibility and Safety Compliance: Ongoing maintenance of pathways, railings, and other safety features is crucial.

- o Solutions: Collaborating with structural engineers and following safety guidelines can prevent accidents.

- Plant Care and Harvesting:
  - o Pruning and Training: Regular care of plants, including pruning and training, optimizes growth and yield.
  - o Seasonal Planning: Ongoing seasonal planning ensures that planting and harvesting align with climate patterns.
  - o Solutions: Implementing a comprehensive planting calendar and training staff or volunteers in plant care techniques.
- Community and Stakeholder Engagement:
  - o Educational Programs: Ongoing education and engagement activities can keep the community involved and supportive.
  - o Solutions: Hosting regular workshops, tours, and community events fosters continued interest and support.

Ongoing care in rooftop farming is a complex but rewarding endeavor that necessitates a blend of horticultural skill, structural awareness, environmental stewardship, and community engagement. From soil fertility to water management, from pest control to safety compliance, and from plant nurturing to community building, every aspect of ongoing care contributes to the resilience, productivity, and vitality of the rooftop farm. It embodies a holistic approach that aligns agricultural practices with urban ecology, social well-being, and economic sustainability. By investing in ongoing care, rooftop farmers not only cultivate crops but also nurture an interconnected ecosystem where urban life and agriculture flourish in harmony.

Energy Efficiency

Energy efficiency in rooftop farming plays a crucial role in minimizing environmental impact and aligning with sustainable urban development goals. Implementing energy-efficient practices can reduce operational costs and contribute to broader energy conservation efforts. Here's an overview of energy efficiency considerations in rooftop farming:

- Solar Energy Utilization:
    - o Solar Panels: Incorporating solar panels into the rooftop farm design can provide renewable energy for lighting, irrigation, and other electrical needs.
    - o Solar Greenhouses: Solar-designed greenhouses can trap heat, reducing the need for artificial heating systems.
    - o Solutions: Collaboration with solar energy experts to design and install appropriate systems.
- Energy-Efficient Water Systems:
    - o Irrigation Techniques: Using energy-efficient irrigation systems, like drip irrigation, can reduce energy consumption related to water pumping.
    - o Water Recycling: Energy-efficient water recycling and rainwater harvesting systems can minimize reliance on external water sources.
    - o Solutions: Regular maintenance and using sensors to optimize water usage.
- Lighting and Climate Control:
    - o LED Lighting: Utilizing energy-efficient LED lights for extended growing seasons or indoor farming spaces.
    - o Insulation and Climate Control: Proper insulation and energy-efficient climate control systems can optimize energy consumption in controlled environments.
    - o Solutions: Timers and sensors to regulate lighting and temperature as needed.
- Materials and Construction:

- Sustainable Materials: Using energy-efficient materials in construction can contribute to overall energy conservation.
  - Energy-Efficient Design: Incorporating energy-saving design principles into the structure of the rooftop farm.
  - Solutions: Consulting with architects and engineers specializing in green construction.

- Transport and Distribution:
  - Local Distribution: Selling produce locally reduces energy costs associated with transportation.
  - Energy-Efficient Vehicles: Utilizing electric or other energy-efficient vehicles for distribution.
  - Solutions: Building connections with local markets and considering alternative distribution methods.
- Monitoring and Optimization:
  - Energy Audits: Regular energy audits can identify areas for improvement in energy consumption.
  - Smart Technology: Implementing smart technology can automate and optimize energy usage.
  - Solutions: Ongoing monitoring and adapting to new energy-efficient technologies as they emerge.

Energy efficiency in rooftop farming is an interdisciplinary challenge that involves technological innovation, thoughtful design, and conscious management practices. From solar energy utilization to energy-efficient water systems, from lighting and climate control to materials and transportation, every aspect of rooftop farming offers opportunities to conserve energy. Such an integrated approach not only aligns with the broader goals of urban sustainability but also makes economic sense, reducing operational costs. The pursuit of energy efficiency in rooftop farming is a continuous journey, reflecting the evolving nature of technology and the dynamic interplay between urban agriculture and the wider urban ecosystem. It exemplifies a commitment to responsible stewardship of resources and a vision of agriculture that nourishes both the city and the planet.

Climate Resilience

Climate resilience in rooftop farming is essential in the face of changing climate patterns and extreme weather events. By adopting strategies that enhance resilience, rooftop farms can continue to thrive and contribute positively to urban environments. Here's an overview of key aspects of climate resilience in rooftop farming:

- Weather-Resilient Infrastructure:
  - Structural Design: Designing rooftop farms to withstand extreme weather conditions, including heavy rains, winds, and temperature fluctuations.
  - Drainage Systems: Implementing efficient drainage systems to manage heavy rainfall and prevent flooding.
  - Solutions: Collaborating with engineers to create weather-resistant structures and systems.
- Water Management:
  - Rainwater Harvesting: Capturing and storing rainwater for irrigation, especially during droughts.
  - Drought-Resistant Crops: Planting crops that require less water can increase resilience during dry periods.
  - Solutions: Integrating water conservation techniques and selecting appropriate crops.
- Temperature Regulation:
  - Shading Solutions: Installing shades or green walls can moderate temperature extremes.
  - Insulation: Using proper insulation to maintain temperature balance during hot and cold weather.
  - Solutions: Utilizing natural and constructed solutions to control temperature.

- Soil and Plant Resilience:
    - Soil Health: Healthy soil can absorb and retain water more efficiently, supporting plants during droughts and heavy rainfall.
    - Diverse Planting: A variety of plant species increases ecological resilience.
    - Solutions: Employing organic farming practices and cultivating diverse plant communities.

- Energy Backup and Efficiency:
    - Renewable Energy Sources: Using solar or wind energy can provide reliable power during energy shortages.
    - Energy Efficiency: Implementing energy-saving practices reduces dependency on external power sources.
    - Solutions: Investing in renewable energy systems and optimizing energy consumption.
- Community Engagement and Education:
    - Local Engagement: Building community awareness and participation in climate resilience strategies.
    - Educational Programs: Providing education on climate resilience and sustainable practices.
    - Solutions: Organizing workshops and fostering community collaboration.

Climate resilience in rooftop farming is a multifaceted approach that requires strategic planning, innovative design, thoughtful management, and community involvement. From weather-resistant infrastructure to water management, from temperature regulation to soil and plant resilience, and from energy backup to community engagement, each aspect contributes to the overall resilience of the rooftop farm. In an era of increasing climate uncertainty, building resilience is not just an adaptive strategy but a proactive commitment to sustainable urban agriculture. By weaving climate resilience into the fabric of rooftop farming, urban farmers not only protect their investments but also contribute to the broader resilience of the urban landscape. It's a path that recognizes the interconnectedness of urban ecosystems and embraces the challenges and opportunities of a changing climate.

Future Directions and Innovation

The exploration of what lies ahead in the world of rooftop farming opens doors to exciting possibilities, offering a glimpse into technological advancements, novel farming practices, enhanced community engagement, and innovative pathways that hold the potential to transform urban sustainability and resilience.

Technological Advancements

Technological advancements in rooftop farming are revolutionizing the way urban agriculture integrates with modern cityscapes. These innovations enhance productivity, sustainability, and adaptability, making rooftop farming more accessible and efficient. Here's an overview of key technological advancements:

- Smart Irrigation Systems:
    - Sensor-Based Technology: Utilizing soil moisture sensors to regulate water usage, ensuring optimal irrigation without waste.
    - Drip Irrigation: Advanced systems that target water directly to the root zone, minimizing evaporation and overwatering.
    - Solutions: Regularly updating and maintaining technology to ensure efficiency.
- Renewable Energy Solutions:
    - Solar Panels: Incorporating solar energy for powering farming operations, reducing reliance on non-renewable sources.

- o Wind Turbines: Small-scale wind energy solutions can also provide a sustainable energy source.
- o Solutions: Collaborating with renewable energy experts to integrate and manage these systems.
- Climate Control Technology:
  - o Automated Greenhouses: Controlled environments with automated temperature, humidity, and light controls to optimize growing conditions.
  - o Energy-Efficient Cooling and Heating Systems: Implementing systems that minimize energy consumption.
  - o Solutions: Regular monitoring and adjustments to keep the systems running effectively.
- Hydroponic and Aquaponic Systems:
  - o Soilless Cultivation: Using nutrient-rich water solutions for growing plants, allowing more control over nutrient balance.
  - o Integrated Fish Farming (Aquaponics): Combining fish farming with plant cultivation in a symbiotic environment.
  - o Solutions: Investing in training and specialized equipment for these advanced growing methods.

- Digital Monitoring and Analytics:
  - o Remote Monitoring: Utilizing software and sensors for real-time monitoring of crop conditions, weather, and other variables.
  - o Data Analytics: Analyzing collected data to make informed decisions about planting, harvesting, and maintenance.
  - o Solutions: Staying abreast of latest software and analytical tools for farming management.
- Robotic Automation:

- o Robotic Planting and Harvesting: Automated systems that can handle repetitive tasks with precision.
- o Drone Surveillance: Drones that can monitor large rooftop farms, providing insights and aerial data.
- o Solutions: Ensuring compatibility with existing farm operations and maintaining the technological infrastructure.

Technological advancements in rooftop farming are shaping the future of urban agriculture, bringing efficiency, sustainability, and adaptability to new heights. From smart irrigation to renewable energy, from climate control to soilless cultivation, from digital analytics to robotic automation, each technological innovation opens new possibilities and challenges. As rooftop farming continues to evolve, embracing these technologies will be key to staying competitive and aligned with sustainability goals. It represents a convergence of agricultural wisdom and technological innovation, a synthesis that reflects the dynamic, forward-thinking nature of urban farming in the 21st century. It's not just about growing food; it's about cultivating a technologically advanced and ecologically responsible approach to urban life.

Scaling Up

Scaling up rooftop farming is a complex yet essential endeavor to make urban agriculture more prominent and viable in modern cities. It involves not only expanding the physical scope but also enhancing efficiency, accessibility, and integration within urban ecosystems. Here's an overview of the critical aspects of scaling up rooftop farming:

- Infrastructure Development:
    - o Structural Assessments: Analyzing existing rooftops and infrastructure to determine suitability for farming.

- Designing Scalable Systems: Creating modular and adaptable designs that can be replicated across various rooftop spaces.
- Solutions: Collaborating with architects, engineers, and urban planners to create scalable designs.

- Financial Models and Investment:
  - Funding Options: Exploring various funding avenues, including grants, private investment, and crowdfunding.
  - Economic Feasibility: Evaluating the financial sustainability and potential return on investment.
  - Solutions: Developing solid business plans and engaging financial experts in the field.

- Regulatory Compliance and Policies:
  - Zoning Laws and Regulations: Navigating city regulations and obtaining necessary permits.
  - Policy Advocacy: Working with policymakers to create supportive laws and incentives.
  - Solutions: Regular consultation with legal experts and active participation in policy development.

- Community Engagement and Partnerships:
  - Local Collaboration: Building partnerships with local businesses, schools, and community organizations.
  - Educational Outreach: Providing education and training to local communities about the benefits of rooftop farming.
  - Solutions: Developing community outreach programs and fostering collaborative relationships.

- Integration with Local Food Systems:
  - Supply Chain Development: Creating reliable supply chains to distribute produce to local markets, restaurants, and stores.
  - Consumer Engagement: Promoting local consumption and awareness of rooftop-grown produce.

- o Solutions: Collaborating with local food stakeholders and marketing directly to consumers.
- Environmental and Social Sustainability:
  - o Sustainable Practices: Ensuring that scaled-up operations adhere to environmentally responsible practices.
  - o Social Impact: Assessing and enhancing the social benefits, including job creation and community well-being.
  - o Solutions: Regularly monitoring and adjusting practices to align with sustainability goals.

Scaling up rooftop farming is a multifaceted challenge that demands a holistic approach. From infrastructure development to financial considerations, from regulatory compliance to community engagement, from integration with local food systems to environmental sustainability, every aspect requires careful planning and execution. It's not merely about enlarging the scale; it's about expanding the vision and impact of rooftop farming within the urban fabric. Scaling up necessitates a synergy of technical expertise, entrepreneurial spirit, regulatory insight, community involvement, and a steadfast commitment to sustainability. The potential rewards are immense, transforming cities into greener, healthier, and more resilient habitats, where rooftop farming becomes an integral part of urban life. It's a bold step towards a future where the cityscape is not just a concrete jungle but a thriving ecosystem, nourished by innovation, collaboration, and a shared vision of urban harmony.

Lessons Learned

As the concept of rooftop farming matures, it has become increasingly evident that the journey is as valuable as the destination. Lessons learned from pioneering projects, successes, failures, and everything in between have begun to shape the future direction of this exciting intersection between agriculture and urban living.

One of the primary lessons learned revolves around the importance of comprehensive planning. Rooftop farming is not simply a matter of placing soil and plants on a roof; it requires careful consideration of structural integrity, water management, climate, and accessibility. Many early projects faced unexpected challenges due to inadequate planning, leading to a greater appreciation for the need for interdisciplinary collaboration involving architects, engineers, agronomists, and other specialists.

Another vital lesson has been the recognition of the necessity for community engagement. Rooftop farms do not exist in isolation; they are part of the broader urban ecosystem and social fabric. Engaging local communities, forging partnerships with local businesses, and integrating with local food systems have emerged as central tenets of successful rooftop farming. These connections enrich the experience and ensure that the farms meet the needs and desires of the community they serve.

Financial sustainability has also proven to be a significant lesson. Many enthusiastic entrepreneurs have embarked on rooftop farming ventures, only to be met with unexpected costs and financial challenges. Finding a viable economic model, securing reliable funding sources, and understanding the potential return on investment has become essential in taking rooftop farming from a niche experiment to a sustainable urban agricultural practice.

Environmental sustainability and social impact have also provided key lessons. Rooftop farming offers exciting opportunities for improving urban ecology, enhancing biodiversity, and contributing to climate resilience. But achieving these goals requires more than good intentions; it requires ongoing commitment, monitoring, and adaptation. Similarly, the social benefits of creating green spaces, educational opportunities, and local employment must be thoughtfully cultivated and nurtured.

Lastly, navigating the regulatory landscape has taught many rooftop farmers valuable lessons about the importance of understanding local laws, zoning regulations, and permitting processes. What might be acceptable in one jurisdiction may be problematic in another. Collaborative efforts with policymakers and active participation in shaping supportive regulations have become crucial strategies.

In conclusion, the lessons learned from rooftop farming are rich and multifaceted. They reveal a landscape that is both exciting and demanding, filled with opportunities and challenges. These lessons provide a roadmap for future endeavors, guiding principles, strategies, and approaches that recognize the complexity and potential of rooftop farming. They reflect a growing maturity in the field, an understanding that rooftop farming is not just an agricultural practice but a complex interplay of engineering, ecology, community, economy, policy, and vision. It's a tapestry woven from many threads, each lesson a strand that contributes to the evolving picture of what urban agriculture can be in the 21st century.

Chapter 3. Community Gardens within Green Infrastructure

Community Gardens within green infrastructure represent a dynamic fusion of urban ecology, social engagement, and sustainable development. As cities around the world grapple with the challenges of urbanization, climate change, and social disconnectedness, community gardens have emerged as multifaceted solutions that resonate with the ideals of green infrastructure. Chapter 3 delves into this exciting interface, exploring how community gardens are not mere spaces for cultivating plants but vibrant ecosystems that contribute to the urban fabric's health and resilience.

This chapter will examine the underlying principles that make community gardens integral to green infrastructure, highlighting their ecological, social, and economic dimensions. From enhancing biodiversity and managing water to fostering community bonds and stimulating local economies, community gardens offer a window into a sustainable urban future. Furthermore, the chapter will explore the collaborative strategies, policy frameworks, and innovative practices that enable community gardens to thrive within the broader context of urban green initiatives.

With case studies, insights from experts, and a holistic perspective, Chapter 3 aims to inspire urban planners, community leaders, policymakers, and citizens alike. Whether you're a seasoned urban gardener or a city dweller curious about green living, this chapter will provide a comprehensive understanding of the role and potential of community gardens within green infrastructure. It is a journey into the heart of urban sustainability, where nature, community, and innovation converge, illuminating a path towards a more harmonious and resilient urban existence.

Introduction to Community Gardens

Community gardens represent a thriving intersection of urban living and agrarian practice, creating shared spaces where people can cultivate not only plants and produce but also community bonds, sustainability, and a deeper connection to the environment.

Connection to Green Infrastructure

Community gardens are integral to green infrastructure in urban environments, offering ecological, social, and economic benefits. Here's how community gardens contribute to green infrastructure:

- Ecological Connection:
  - Biodiversity Enhancement: Providing habitats for diverse plant species and local wildlife, including pollinators.
  - Climate Mitigation: Contributing to carbon sequestration and reducing heat island effects.
  - Water Management: Utilizing rainwater harvesting and permeable surfaces to reduce stormwater runoff.
  - Solutions: Collaborating with ecologists to maximize ecological benefits and sustainability.
- Social Integration:
  - Community Building: Creating shared spaces that foster community interaction, cooperation, and a sense of belonging.
  - Education and Awareness: Offering educational opportunities for sustainable living, organic farming, and environmental stewardship.
  - Health and Well-being: Encouraging outdoor activities, healthy eating, and mental well-being through gardening.

- o Solutions: Engaging with local community organizations and educational institutions to enhance social integration.
- Economic Implications:
  - o Local Food Production: Reducing food transportation costs and supporting local economies through community-supported agriculture.
  - o Job Creation: Providing opportunities for employment in gardening, maintenance, and education.
  - o Solutions: Partnering with local businesses and government to create sustainable economic models.

- Connection to Wider Green Infrastructure:
  - o Integration with Parks and Greenways: Forming part of a broader network of green spaces, enhancing overall urban green infrastructure.
  - o Collaboration with Other Urban Agriculture Initiatives: Linking with rooftop farms, urban orchards, and other green initiatives to create a cohesive urban agriculture landscape.
  - o Solutions: Working with urban planners and landscape architects to ensure seamless integration within the urban fabric.
- Policy and Regulation:
  - o Supportive Legislation: Advocating for and benefiting from policies that support community gardens and urban agriculture.
  - o Compliance with Local Regulations: Ensuring alignment with zoning laws, land use, and other local regulations.
  - o Solutions: Regularly liaising with policymakers and legal experts to ensure compliance and advocacy.

Community gardens are more than just spaces for growing food; they are vital components of urban green infrastructure, providing a multifaceted contribution to city life. From enhancing biodiversity to fostering community, from supporting the local economy to integrating within the broader green infrastructure network, community gardens embody a sustainable and resilient urban vision. Through thoughtful planning, collaboration, and alignment with broader urban goals, community gardens can flourish as living examples of green innovation, social cohesion, and responsible urban development. The connection between community gardens and green infrastructure represents a harmonious blend of nature, community, and sustainability, painting a vibrant picture of what urban living can aspire to be.

Benefits to Communities

Community gardens offer an array of benefits to urban communities that extend far beyond the mere cultivation of plants. These shared spaces foster a sense of connection, wellbeing, and shared responsibility, contributing to the overall quality of urban life. Below, we explore the manifold benefits community gardens bring to communities:

- Social Cohesion and Community Building:
  - Fostering Relationships: Community gardens provide common grounds for people from diverse backgrounds to come together, collaborate, and form lasting bonds.
  - Building Trust and Cooperation: Shared responsibilities and common goals help foster a sense of trust and cooperation among community members.
- Health and Wellness:
  - Physical Health: Engaging in gardening activities offers physical exercise, exposure to fresh air, and access to fresh, nutritious produce.

- o   Mental Wellbeing: Gardens offer a therapeutic escape from urban hustle, reducing stress, and enhancing mental health.
- Education and Skill Development:
  - o   Hands-on Learning: Gardens serve as outdoor classrooms where individuals of all ages can learn about horticulture, nutrition, and environmental stewardship.
  - o   Skill Enhancement: Community gardens provide opportunities to learn various skills such as composting, water management, and organic farming techniques.
- Environmental Stewardship:
  - o   Sustainability Education: By practicing sustainable gardening methods, community gardens educate members about ecological practices.
  - o   Biodiversity Enhancement: These spaces can be designed to support local flora and fauna, enriching urban biodiversity.
- Economic Benefits:
  - o   Access to Fresh Produce: Community gardens provide affordable access to fresh, locally grown produce, enhancing food security.
  - o   Job Opportunities: Gardens can create employment opportunities in garden management, education, and related services.
- Cultural Expression and Heritage Preservation:
  - o   Cultural Exchange: Gardens can be places where cultural traditions are shared and celebrated through the cultivation of heritage plants and communal events.
  - o   Preservation of Traditional Knowledge: They can act as repositories for traditional gardening knowledge and techniques, preserving cultural heritage.

Community gardens are multifaceted assets to urban communities, enriching the lives of those involved in numerous ways. They provide a bridge between urban living and nature, between individuals and community, and between the present and a sustainable future. The benefits they bring are not just individual but collective, weaving together social, economic, environmental, and cultural threads to create a resilient and vibrant urban tapestry. Whether it's a small neighborhood plot or a large urban farm, community gardens symbolize a progressive shift in how urban communities perceive and interact with their environment, nurturing not only plants but the very essence of community itself.

Local Examples

Community gardens, with their unique blend of social, environmental, and cultural benefits, are increasingly becoming cornerstones of urban development. Local examples of community gardens reflect the diverse ways in which these spaces can be leveraged to enhance community well-being.

In New York City, community gardens like La Plaza Cultural have transformed abandoned lots into thriving green spaces, creating a refuge for both nature and community engagement. These gardens are not only places to grow food but have become centers for artistic expression, cultural events, and environmental education.

San Francisco's Alemany Farm illustrates how a community garden can be a catalyst for social change. Located in a neighborhood with limited access to fresh produce, the farm addresses food security while offering hands-on education about sustainable agriculture.

Across the Atlantic in London, the Incredible Edible project is turning public spaces into communal vegetable gardens. The initiative fosters community spirit and provides free access to fresh produce, all while beautifying the urban landscape.

In Tokyo, Japan, Machimura Farm creates opportunities for urban residents to reconnect with agriculture. Through its community gardens and educational programs, the project promotes awareness of sustainable farming practices and fosters a sense of stewardship over the land.

In Cape Town, South Africa, the Philippi Horticultural Area highlights the role community gardens can play in economic empowerment. By providing training and resources, the project supports local farmers, enhancing livelihoods, and contributing to community resilience.

In Sydney, Australia, the community gardens movement has brought together people from various cultural backgrounds. Gardens like the Callan Park Community Garden facilitate social inclusion, promote cross-cultural understanding, and create a sense of belonging among diverse community members.

Toronto's Regent Park Community Garden in Canada is an example of how urban gardens can be integrated into city planning. This garden is part of a broader revitalization effort, improving the quality of life in the neighborhood while promoting urban sustainability.

These local examples underscore the universality of community gardens. From addressing food scarcity to promoting social cohesion, from beautifying urban spaces to enhancing mental and physical well-being, community gardens are demonstrating their transformative potential. The flexibility, creativity, and community-driven approach inherent in these gardens make them adaptable to various urban contexts. They are local solutions with global resonance, reflecting a growing recognition of the essential role community gardens can play in building a sustainable, healthy, and harmonious urban future.

Planning and Development

Laying the groundwork for successful community farming requires a multifaceted approach that encompasses careful planning, collaboration with local stakeholders, adherence to regulatory guidelines, and a vision that aligns with the broader goals of sustainability and community well-being.

Site Selection within Urban Fabric

Site selection for community gardens within the urban fabric is a critical step that requires careful consideration of various factors. This process not only determines the success of the garden but also its impact on the broader community and urban environment. Below are key considerations and aspects involved in site selection:

1. Location Accessibility: Community gardens should be easily accessible to those who will use and benefit from them. Proximity to residential areas, availability of public transportation, and safe walking routes are essential in ensuring that a broad spectrum of community members can access the garden.
2. Land Suitability: The physical characteristics of the site, including soil quality, sunlight exposure, and water availability, must be conducive to gardening. Conducting soil tests and evaluating the site's potential for cultivation is vital to avoid future challenges.
3. Legal Considerations: Acquiring the necessary permissions and ensuring that the land's use aligns with zoning regulations is a critical aspect of site selection. Engaging with local authorities to understand legal constraints and opportunities can smooth the process.

4. Community Needs and Involvement: Engaging with the community to assess their needs and interests ensures that the garden reflects local desires and needs. Community participation in site selection fosters ownership and long-term engagement with the garden.

5. Environmental Impact: Assessing and minimizing the potential environmental impact of the garden is crucial. This includes considering aspects such as water management, sustainable design practices, and the potential effect on local ecosystems.

6. Integration with Existing Urban Infrastructure: The garden should seamlessly integrate with existing urban structures and landscapes. Understanding the broader urban context helps in creating a garden that complements and enhances the surrounding area.

7. Economic Considerations: Evaluating the costs associated with site preparation, maintenance, and potential funding sources ensures that the garden is economically viable. Partnerships with local businesses, grants, and community support can provide financial sustainability.

8. Long-term Sustainability: The site selection process must consider the long-term sustainability of the garden. This involves assessing the availability of resources, community support, and the potential for future growth or adaptation.

9. Cultural and Historical Context: Recognizing and respecting the cultural and historical significance of potential sites can enhance community engagement and avoid potential conflicts.

10. Potential for Multi-Functionality: Sites that can accommodate various functions such as educational programs, social gatherings, or therapeutic activities add additional value to the community garden.

In conclusion, site selection within the urban fabric is a complex process that demands a multifaceted approach. By considering the physical, social, legal, environmental, and economic aspects, community gardens can become valuable assets that resonate with the local community's needs and aspirations. It's a delicate balance that requires collaboration, foresight, and a clear understanding of the urban context and the transformative potential of community gardens.

Designing for Sustainability

Designing for sustainability in community gardens and urban agriculture is a multifaceted process that emphasizes responsible resource management, ecological balance, and long-term viability. Here's how designers, urban planners, and community members can approach this essential task:

1. Water Conservation: Utilizing techniques such as rainwater harvesting, drip irrigation, and xeriscaping (landscaping with drought-tolerant plants) can significantly reduce water consumption. These practices not only preserve a vital resource but also contribute to managing stormwater runoff.
2. Soil Health Management: Employing organic gardening practices like composting, cover cropping, and no-till farming helps maintain soil health. These techniques enhance soil structure, increase water retention, and promote beneficial microbial activity.
3. Energy Efficiency: The use of renewable energy sources, such as solar or wind power for lighting or irrigation, minimizes energy consumption. Selecting energy-efficient equipment and using passive solar design principles can further contribute to energy conservation.
4. Recycling and Waste Reduction: Encouraging the use of recycled and locally-sourced materials for garden structures and promoting on-site composting of organic waste can significantly reduce the ecological footprint.

5. Biodiversity Protection: Designing with native plants and creating habitats for local wildlife enhances biodiversity. Planting diverse crops also aids in pest management by reducing the need for chemical interventions.

6. Community Involvement: Engaging the community in the design process ensures that the garden meets local needs and encourages stewardship. Education about sustainable practices also fosters a culture of responsibility and ongoing care.

7. Economic Sustainability: Developing models that support local economies, such as community-supported agriculture (CSA) or farmers' markets, ensures economic sustainability. Funding strategies and partnerships can be explored to support initial development and ongoing maintenance.

8. Accessibility and Inclusivity: Designing for all age groups and abilities promotes inclusivity. Features such as raised beds, wide paths, and appropriate signage can make the garden accessible to everyone.

9. Resilience to Climate Change: Climate-adaptive design includes selecting plants that are tolerant to local weather conditions and implementing water management strategies that account for changing precipitation patterns.

10. Holistic Integration with Urban Environment: The garden's design should consider its relationship with the surrounding urban fabric. This includes aesthetic considerations, functional connections with other urban green spaces, and alignment with broader urban sustainability goals.

In summary, designing for sustainability in community gardens is a dynamic and comprehensive task that involves intertwining ecological, social, economic, and cultural considerations. It demands a collaborative and thoughtful approach, one that recognizes the garden as part of a broader ecosystem. When done effectively, sustainable design transforms community gardens into vibrant, resilient spaces that nourish communities, enrich urban landscapes, and contribute to a sustainable future for all.

Tools and Resources

In the development and maintenance of community gardens, rooftop farms, or other urban agriculture initiatives, the right tools and resources are crucial. They not only facilitate the physical creation and care of these spaces but also contribute to their sustainability, community engagement, and educational outreach. Here's an overview of the essential tools and resources:

- Physical Tools:
    - Gardening Tools: Shovels, rakes, hoes, pruners, and other hand tools are fundamental for planting, maintenance, and harvesting.
    - Irrigation Systems: Tools and equipment for efficient water management, including hoses, watering cans, drip irrigation systems, etc.
    - Composting Facilities: Compost bins or tumblers to manage organic waste and improve soil quality.
    - Safety Equipment: Gloves, goggles, and other personal protective equipment for safe gardening practices.
- Technological Tools:
    - Software for Design and Planning: Tools that help in visualizing and planning garden layouts, irrigation systems, or solar energy utilization.
    - Weather and Soil Monitoring Devices: Gadgets that provide real-time information on weather conditions and soil health to optimize planting and watering schedules.
    - Online Platforms for Community Engagement: Websites, social media, or mobile apps that facilitate

communication, scheduling, and community involvement.

- Educational Resources:
    - Workshops and Training Programs: Offering classes on gardening techniques, sustainability practices, cooking with garden produce, etc.
    - Educational Materials: Books, brochures, or online content that provide information on various aspects of urban agriculture.
    - School and Community Partnerships: Collaborations with local schools, universities, and community organizations to promote learning and community development.
- Community and Social Resources:
    - Community Engagement Tools: Strategies and platforms to involve community members in decision-making, volunteering, and ongoing garden care.
    - Social Support Networks: Creating networks with other community gardens, local businesses, or governmental bodies for mutual support and resource sharing.

- Financial Resources:
    - Funding and Grants: Information and access to governmental grants, private funding, or crowdfunding platforms to support the initial setup and ongoing operations.
    - Business Models: Developing sustainable economic models, such as CSA programs or farmers' markets, to generate revenue and ensure financial stability.
- Legal and Regulatory Resources:
    - Zoning and Land Use Information: Guidance on local zoning laws, land-use regulations, and obtaining necessary permits.

o   Liability and Insurance Considerations: Understanding and managing legal responsibilities and obtaining appropriate insurance coverage.

In conclusion, a combination of physical tools, technological advancements, educational offerings, community engagement strategies, financial support, and legal guidance forms the foundational resources for successful urban agriculture endeavors. These tools and resources are interdependent and collectively contribute to creating vibrant, sustainable, and enriching urban green spaces.

Community Involvement and Education

Fostering a sense of ownership and connection within the community, paired with ongoing education and awareness-building, can transform a community garden from a simple plot of land into a vibrant hub of learning, collaboration, and empowerment.

Building Community Connections

Building community connections is at the core of successful urban agriculture and green infrastructure initiatives. Community gardens, rooftop farms, and other urban green spaces are not just places for growing food and enhancing biodiversity; they are platforms for fostering social ties, civic engagement, and a sense of communal identity.

The inception of an urban agriculture initiative often begins with identifying the needs and aspirations of the community it intends to serve. Community meetings, surveys, and outreach events provide opportunities to understand local desires and to involve residents in the planning and decision-making processes. By actively engaging community members in the design and management of urban green spaces, planners and organizers create a sense of ownership and responsibility that can be vital for long-term success.

Urban agriculture initiatives often act as social hubs, where neighbors meet and forge relationships. Through collaborative planting, maintenance, and harvesting activities, diverse community members – regardless of age, ethnicity, or social background – find common ground and develop mutual respect. These interactions break down social barriers, build trust, and foster an environment of support and collaboration.

Educational programs within urban agriculture spaces can further contribute to community connections. Workshops on gardening techniques, environmental stewardship, or healthy cooking can attract various community members, including schools, families, and seniors. These educational experiences not only impart knowledge but create opportunities for intergenerational exchange and shared learning.

Collaboration with other community organizations, local businesses, and governmental bodies can broaden the reach and impact of urban agriculture. Partnerships can result in shared resources, expert guidance, and strengthened community networks. Collaborative events, like farmers' markets or community festivals, can become cultural landmarks that celebrate local identity and traditions.

The economic aspect of urban agriculture can also contribute to community connections. Community-supported agriculture (CSA) programs, for example, create direct links between growers and consumers. They promote local economic development and foster a sense of community investment in local food production.

However, building community connections isn't without challenges. Effective communication, inclusivity, conflict resolution, and sustained engagement require careful consideration and ongoing effort. Leaders and organizers must be attentive to the diverse needs and voices within the community, ensuring that the initiative remains responsive, inclusive, and adaptive.

In conclusion, urban agriculture is more than the mere cultivation of crops within city limits; it's a dynamic social endeavor that cultivates community connections. By promoting collaboration, inclusivity, education, and local economic growth, urban green spaces become vital community assets that enrich urban life, strengthen social fabric, and contribute to the overall well-being and resilience of the community.

Environmental Education

Environmental education within the specific context of community gardens takes on a unique and pivotal role, providing opportunities for communities to engage directly with nature and sustainable practices right in their urban surroundings.

Community gardens serve as living laboratories where environmental education comes to life. These gardens offer a space for hands-on experiences, where individuals can learn about soil health, water conservation, composting, pest management, and other sustainable gardening practices. The tactile nature of gardening allows participants to witness ecological principles in action, from seed germination to plant-pollinator interactions.

Schools partnering with community gardens can enrich their curriculum by aligning garden activities with science, mathematics, and social studies lessons. Students can explore topics such as plant biology, weather patterns, and even local history within the garden. These real-world applications enhance understanding and retention, fostering a connection to the environment that transcends traditional classroom boundaries.

For adults and lifelong learners, community gardens offer workshops, demonstrations, and volunteer opportunities that empower them with practical knowledge and skills. Whether it's mastering the art of composting or learning about native plant species, these educational experiences cater to various interests and needs within the community.

Beyond technical gardening skills, community gardens foster awareness of broader environmental and social issues. They provide a platform for discussions about food security, local food systems, climate change, and urban sustainability. Participants can explore these complex topics in a tangible way, as they see how their individual and collective efforts contribute to the larger ecological and societal goals.

Community gardens also promote environmental stewardship and civic engagement. Through involvement in garden planning, maintenance, and governance, individuals take on roles of responsibility and leadership within their community. They become advocates for green spaces, sustainable practices, and policies that reflect environmental consciousness.

In addition, community gardens facilitate social connections that enhance environmental learning. The communal nature of these spaces encourages collaboration, dialogue, and mutual respect among diverse members. This social interaction enriches the educational experience, creating a supportive learning environment where knowledge, ideas, and values are shared and cultivated.

Environmental education in community gardens is not a one-size-fits-all approach. It requires attention to the specific needs, interests, and characteristics of the community. Collaborating with local organizations, universities, and experts can tailor educational programs to reflect the local culture, ecology, and goals.

In conclusion, community gardens are vibrant hubs for environmental education within urban landscapes. They transform abstract environmental concepts into concrete, experiential learning. By connecting people with nature and each other, community gardens foster environmental literacy, stewardship, and a shared commitment to sustainability, enhancing the overall well-being and resilience of the community.

Health and Well-being

Community gardens contribute significantly to the health and well-being of urban populations, offering multifaceted benefits that encompass physical, mental, and social dimensions.

Physically, community gardens promote a healthy lifestyle by encouraging physical activity. Tending to a garden involves various forms of exercise, from digging and planting to watering and weeding. These activities can contribute to overall physical fitness and reduce the risk of chronic diseases such as obesity and heart problems. The gardens also provide access to fresh, locally grown fruits and vegetables, which can improve dietary habits and nutritional intake.

Mentally, the act of gardening has been shown to reduce stress, anxiety, and depression. Connecting with nature, nurturing plants, and watching them grow provide therapeutic effects that calm the mind and elevate the mood. The repetitive tasks in gardening can create a meditative state that enhances mindfulness and emotional well-being. In this way, community gardens can serve as a respite from urban hustle and bustle, offering a place for reflection and relaxation.

Socially, community gardens act as communal spaces where neighbors come together, share experiences, and build relationships. They foster a sense of community belonging and connection, breaking down social barriers and creating a support network. These interactions can enhance social well-being by providing opportunities for friendship, collaboration, and a shared sense of purpose.

Furthermore, community gardens play a role in enhancing psychological well-being through empowerment and self-efficacy. Participating in a garden project from planning to harvesting gives individuals a sense of accomplishment and control over their environment. It can boost self-esteem and provide a sense of purpose, particularly for those who may feel marginalized or isolated in their urban settings.

Community gardens can also serve specific therapeutic purposes. Horticultural therapy programs within these gardens can assist individuals with disabilities, mental health issues, or rehabilitation needs. Guided gardening activities, tailored to individual abilities and goals, provide therapeutic benefits that support recovery and enhance the quality of life.

Health and well-being in the context of community gardens extend beyond individual benefits. They contribute to community resilience and cohesion by promoting social inclusion, community engagement, and environmental stewardship. The gardens become symbols of community identity and pride, reflecting shared values and aspirations for a healthier and more sustainable urban life.

In summary, community gardens are more than just places to grow food; they are nurturing environments that enhance health and well-being on multiple levels. Through physical engagement with the soil, therapeutic connections with nature, social interactions, and empowerment, community gardens contribute to a holistic vision of health that resonates with individuals and communities alike, making them valuable assets in urban health promotion and community development.

Legal and Regulatory Considerations

Navigating the complex web of laws, regulations, zoning requirements, and municipal codes is a critical step in establishing and sustaining urban agriculture projects, requiring thorough understanding and careful alignment with both local and national legal frameworks.

Land Access and Ownership

Land access and ownership within the realm of community gardens and urban agriculture present both opportunities and challenges that require careful consideration and strategic planning.

In densely populated urban areas, access to suitable land for gardening and farming can be a significant obstacle. Limited availability of open space, competition with commercial and residential developments, and high land prices often restrict opportunities for community garden establishment. Identifying and securing suitable land thus becomes a pivotal task that demands collaboration between local authorities, community organizations, landowners, and other stakeholders.

Ownership structures for community gardens vary widely and can include public ownership, private ownership, leasing, or cooperative arrangements. Each of these models carries its advantages and potential challenges.

Public ownership, where the local government owns and supports the garden, can provide stability and long-term security. However, it may also entail bureaucratic processes that hinder flexibility and community autonomy.

Private ownership, on the other hand, can offer more control and adaptability but may lack long-term security, especially if the landowner decides to sell or repurpose the land. Leasing arrangements can provide temporary solutions but might also limit investment in long-term infrastructure and improvements.

Community land trusts and cooperative ownership models present innovative solutions that emphasize community control and shared responsibility. These models can ensure long-term stability and reflect community needs and values, but they often require significant legal, financial, and organizational resources to establish and maintain.

Local zoning regulations, land use policies, and legal frameworks also play a crucial role in shaping land access and ownership for community gardens. Proactive policies that recognize community gardens as valuable urban land use can facilitate access to land, provide legal protection, and offer incentives such as tax benefits or grants.

Successful land access and ownership strategies typically involve a multifaceted approach that integrates community engagement, legal expertise, financial planning, and partnership building. Collaborating with local government, nonprofit organizations, legal experts, and other community stakeholders can create tailored solutions that align with community needs and aspirations.

Overall, land access and ownership are fundamental to the success and sustainability of community gardens. Navigating this complex landscape requires thoughtful planning, flexibility, collaboration, and a clear understanding of the community's goals and values. Creative solutions that honor community participation and foster long-term stability can transform urban vacant lots into thriving gardens that enrich the urban fabric, providing spaces for growth, connection, and sustainability.

Regulations and Incentives

Community garden regulations and incentives play a crucial role in the development, sustainability, and success of urban gardening projects. Understanding these elements is essential for community garden organizers and participants.

Regulations

- Zoning Laws:
  - Define where community gardens can be located.
  - May restrict certain types of gardening activities, such as raising livestock.
  - Can vary significantly between cities and municipalities.
- Land Use Agreements:
  - Outline the rights and responsibilities of gardeners and landowners.
  - Include terms regarding maintenance, liability, and potential conflicts.
- Water Usage Regulations:
  - Govern the use of water for irrigation.
  - May restrict or limit water usage, particularly in drought-prone areas.
- Health and Safety Codes:

- o Ensure that community gardens comply with sanitation, accessibility, and safety standards.
  - o Might cover aspects like composting practices, pesticide usage, and tool storage.
- Building Codes and Permits:
  - o Apply to structures like sheds, greenhouses, or fences within the garden.
  - o Require compliance with local construction standards and permitting processes.

Incentives

- Grants and Funding Opportunities:
  - o Provided by governmental agencies, private foundations, or corporations.
  - o Support the start-up, expansion, or enhancement of community gardens.
- Tax Incentives and Abatements:
  - o Offered to landowners who make their property available for community gardening.
  - o Can reduce property taxes or provide other financial benefits.
- Technical Assistance and Resources:
  - o Include support from agricultural extension services, universities, or nonprofits.
  - o Offer expertise in garden design, soil testing, education, and community organizing.
- Public Recognition and Awards:
  - o Acknowledge and celebrate successful community garden projects.
  - o Enhance the visibility and reputation of the garden within the community.

The complex landscape of regulations and incentives requires community garden organizers to engage with local authorities, legal experts, and other stakeholders. Understanding and navigating these rules and opportunities can ensure compliance, enhance sustainability, and leverage support for community garden initiatives. It reinforces the critical role that community gardens play in promoting urban agriculture, environmental stewardship, social cohesion, and public health, reflecting a broader societal commitment to sustainable urban development.

Challenges and Solutions

Community gardens present several challenges, but with thoughtful planning and collaboration, solutions can be found to overcome these obstacles. Here's an overview of common challenges and potential solutions:

Challenges

- Land Accessibility:
    - Difficulty in Finding Suitable Land: Urban areas often lack available green space.
    - High Costs: Land prices can be prohibitive.
- Legal and Regulatory Hurdles:
    - Complex Zoning Laws: Regulations may restrict garden locations or activities.
    - Liability Issues: Concerns over accidents or disputes between gardeners and landowners.
- Resource Constraints:
    - Limited Funding: Startup and maintenance costs may be substantial.
    - Water Availability: Access to affordable water sources can be a challenge.
- Community Engagement:
    - Diverse Interests: Balancing the needs and wants of different community members.

- o Lack of Participation: Sustaining volunteer involvement and commitment.
- Environmental Concerns:
  - o Soil Contamination: Urban soil may contain pollutants or toxins.
  - o Pest Management: Controlling pests without harmful chemicals.

Solutions

- Land Accessibility:
  - o Partnerships with Landowners: Working with private or public landowners.
  - o Use of Vacant Lots: Temporary use of unused urban spaces.
- Legal and Regulatory Solutions:
  - o Clear Agreements: Drafting contracts that define responsibilities.
  - o Engage with Local Authorities: Collaborate to understand and comply with regulations.
- Resource Solutions:
  - o Fundraising and Grants: Seek support from local businesses, grants, and community fundraising.
  - o Water Conservation Techniques: Implement rain barrels, drip irrigation.
- Community Engagement Solutions:
  - o Inclusive Planning: Involve community members in design and decision-making.
  - o Regular Communication: Keep volunteers informed and engaged through meetings and newsletters.
- Environmental Solutions:
  - o Soil Testing: Before planting, test soil for contaminants.
  - o Organic Practices: Use organic methods to manage pests and improve soil health.

In conclusion, while community gardens face many challenges, a combination of creativity, collaboration, planning, and persistence can yield effective solutions. Engaging with community members, local authorities, experts, and stakeholders can lead to a shared vision and practical strategies that transform these challenges into opportunities. The result is a thriving community garden that not only produces fresh food but fosters social connections, enhances well-being, and contributes to a more sustainable urban landscape.

Long-term Management and Impact

Ensuring the success and positive impact of urban agriculture projects over time necessitates a strategic approach to long-term management, focusing on adaptability, sustainability, ongoing community engagement, and continuous evaluation of environmental, social, and economic outcomes.

Sustainability Goals

Community gardens are emerging as essential elements in achieving sustainability goals within urban areas. They contribute to a more sustainable future in various ways, each aligned with broader objectives of environmental stewardship, social equity, and economic viability.

First and foremost, community gardens promote environmental sustainability by maximizing the use of urban spaces for greenery and food production. These gardens help in reducing food miles, cutting down on transportation-related emissions, and providing locally sourced, fresh produce. They also enhance biodiversity by supporting various plant species and beneficial insects. Techniques like composting, organic farming, and water conservation within the garden contribute to efficient resource management, minimizing waste, and reducing the overall ecological footprint.

Community gardens also contribute to social sustainability by fostering connections among diverse members of urban communities. They serve as gathering spaces where people from different backgrounds, ages, and cultures can collaborate, share knowledge, and build friendships. The hands-on experience of growing food helps reconnect urban residents with nature, enhancing awareness of environmental issues and promoting a sense of stewardship for the planet. Additionally, community gardens can play a role in improving public health by providing access to nutritious fresh food and offering opportunities for physical activity.

Economic sustainability is another key benefit of community gardens. They can provide low-cost access to fresh produce for economically disadvantaged community members, promoting food security and reducing dependence on external food systems. Some community gardens also develop into micro-enterprises, where surplus produce is sold at local markets, generating income for gardeners and supporting local economies.

In the context of broader urban planning and development, community gardens align with sustainability goals by creating resilient, adaptive spaces that can respond to changing climate conditions. They offer opportunities for urban heat island mitigation, stormwater management, and carbon sequestration, contributing to the overall resilience of the urban fabric.

However, the successful integration of community gardens into urban sustainability goals requires thoughtful planning, collaboration, and ongoing support. Partnerships between local governments, community organizations, residents, and other stakeholders are vital to navigate the challenges of land access, funding, regulations, and community engagement.

Overall, community gardens are more than just spaces for growing food; they are dynamic, multifaceted platforms that contribute to the realization of sustainability goals. Through their impact on environmental, social, and economic domains, they exemplify the integrated approach needed to create vibrant, sustainable urban communities for the future.

Monitoring and Evaluation

Monitoring and evaluation (M&E) in community gardens are essential for ensuring that garden projects meet their objectives, adhere to best practices, and continuously improve. Here's an overview of key aspects related to monitoring and evaluation in community gardens:

Monitoring

Monitoring is an ongoing process that tracks progress and performance in community gardens:

1. Goals and Objectives:
    o Establish clear, measurable goals for the garden.
    o Include both short-term and long-term objectives, such as increasing food production, enhancing community engagement, or improving environmental sustainability.
2. Data Collection:
    o Regularly collect data on key indicators, such as crop yields, volunteer hours, community participation, water usage, etc.
    o Use tools like logbooks, surveys, observations, or digital tracking systems.
3. Progress Tracking:
    o Monitor the progress towards achieving the defined goals.

- o Analyze trends, identify challenges, and adapt strategies as needed.
4. Quality Assurance:
   - o Ensure that garden practices adhere to quality standards, such as organic farming methods or safety protocols.
   - o Implement regular inspections and feedback loops with garden participants.

Evaluation

Evaluation is a more formal assessment process that examines the effectiveness, efficiency, and relevance of the community garden in achieving its goals:

1. Impact Assessment:
   - o Assess the garden's broader impact on the community, including social, environmental, and economic effects.
   - o Utilize methods like interviews, focus groups, or comparative studies.
2. Cost-Effectiveness Analysis:
   - o Evaluate the financial aspects, including budgeting, spending, and return on investment.
   - o Consider the value of non-monetary benefits, such as social cohesion or health improvements.
3. Stakeholder Engagement:
   - o Involve community members, volunteers, partners, and funders in the evaluation process.
   - o Collect feedback and incorporate diverse perspectives to ensure a comprehensive analysis.
4. Continuous Improvement:
   - o Use evaluation findings to make informed decisions, improve practices, and enhance outcomes.
   - o Implement lessons learned and best practices in future planning and implementation.

M&E in community gardens are vital processes that enable garden organizers and stakeholders to understand how well the garden is functioning and where improvements can be made. By systematically collecting data, analyzing progress, evaluating impact, and engaging with community members, M&E offers valuable insights and actionable recommendations. It fosters a culture of accountability, transparency, and continuous learning that is essential for the long-term success and sustainability of community garden initiatives.

Future Prospects

The future prospects for community gardens are remarkably bright and diverse, as they continue to evolve and respond to the multifaceted needs of urban communities. Positioned at the intersection of urban agriculture, environmental stewardship, community building, and health promotion, community gardens are uniquely capable of addressing many of the complex challenges facing modern cities.

Firstly, community gardens offer a robust solution to food security challenges, particularly in underserved urban areas. As the global population continues to urbanize and the demand for local, fresh produce increases, community gardens could play a crucial role in creating resilient and decentralized food systems. By empowering local residents to grow their food, they contribute to reduced reliance on long supply chains and promote healthier dietary choices.

Furthermore, community gardens are emerging as vital tools in combating climate change and enhancing environmental sustainability. They act as carbon sinks, stormwater management systems, and urban green spaces, all of which are integral in building climate-resilient cities. The emphasis on organic practices, biodiversity, and resource conservation in community gardening aligns with broader environmental goals, making them relevant in the context of global sustainability agendas.

Socially, community gardens foster community cohesion and empowerment. They act as spaces for learning, collaboration, and cultural exchange, strengthening social bonds and creating inclusive environments. The future could see community gardens becoming integral parts of educational curricula, promoting hands-on environmental education and civic engagement from a young age.

Economically, the potential of community gardens to generate income, create jobs, and support local economies is increasingly recognized. Whether through selling surplus produce or developing garden-related enterprises, community gardens contribute to economic empowerment, particularly for marginalized populations.

The future of community gardens is also closely tied to technological advancements and innovations. From smart irrigation systems to data-driven monitoring tools, technology could enhance the efficiency, productivity, and sustainability of community gardens. Integrating these advancements with traditional gardening wisdom offers an exciting pathway for future growth.

However, realizing these future prospects requires concerted efforts across policy, planning, funding, and community engagement. Addressing challenges such as land access, regulatory barriers, and sustainable funding models is essential. Collaboration between governments, NGOs, private sectors, and local communities will be key in leveraging the full potential of community gardens.

Overall, community gardens present an optimistic future, holding the promise of transforming urban landscapes into more sustainable, resilient, and inclusive spaces. Their multifunctional nature and adaptability position them as critical assets in shaping the future of urban living, a future rooted in community, sustainability, and well-being.

Chapter 4. Case Studies

Chapter 4 presents an in-depth examination of urban agriculture through the lens of three distinct and vibrant cities: Detroit, Berlin, and Singapore. Each of these cities has embraced urban agriculture in unique ways, reflecting diverse cultural, environmental, economic, and social contexts.

In Detroit, a city known for its industrial past and urban revitalization, community gardens and urban farms have become symbols of hope and regeneration. The transformation of vacant lots into productive spaces not only fosters community cohesion but also stimulates local economies.

Berlin, with its rich history and progressive urban planning, offers a mix of rooftop farms, community gardens, and innovative agricultural practices. The city's focus on sustainability and citizen participation has created a dynamic urban agriculture scene that integrates seamlessly with its green infrastructure goals.

Singapore, a bustling city-state with limited land, has turned to vertical farming and other innovative solutions to address its food security challenges. With a strong governmental push and cutting-edge technology, Singapore stands as a leading example of how urban agriculture can be implemented in densely populated areas.

Together, these case studies offer valuable insights into the varied approaches to urban agriculture, highlighting successes, challenges, and lessons learned. They provide a comprehensive understanding of how urban agriculture can be adapted to different urban landscapes, and they underscore the importance of collaboration, innovation, and community engagement.

By examining these three cases, readers will gain a deeper appreciation of the multifaceted nature of urban agriculture, its potential to transform urban spaces, and its role in advancing sustainability and resilience. Whether a policymaker, urban planner, researcher, or urban agriculture enthusiast, this chapter provides real-world examples that inspire and inform, paving the way for future urban agricultural endeavors across diverse cityscapes.

Case Studies

Numerous cities around the world have implemented urban agricultural initiatives, including the following.

Detroit, USA: Urban Agriculture as a Catalyst for Change

The decline of the automotive industry in Detroit left the city with vast areas of vacant land and a series of socio-economic challenges. The emergence of urban agriculture has become a transformative force, providing innovative solutions to these problems.

- Urban Revitalization:
    - Reclaiming Abandoned Land: With an estimated 23 square miles of vacant land, various initiatives have turned these lots into productive farms and gardens.
    - Beautification: Urban farming has contributed to the aesthetic improvement of neighborhoods, turning derelict spaces into vibrant, green areas.
    - Property Value Enhancement: Proximity to urban gardens has shown an increase in property values, stimulating further investment in these areas.
- Community Engagement and Empowerment:

- o Community Gardens: Organizations like Keep Growing Detroit support over 1,400 urban gardens, fostering a sense of community ownership.
- o Education and Training: Programs provide hands-on learning opportunities for adults and children, promoting agricultural skills and healthy eating.
- o Social Integration: Urban agriculture has become a unifying force, bringing together diverse community members and fostering social cohesion.
- Food Security:
  - o Addressing Food Deserts: Many areas of Detroit lack access to fresh produce. Urban agriculture is bridging this gap, providing locally grown, affordable food.
  - o Farmers' Markets: Markets like Eastern Market offer fresh produce, strengthening the connection between urban farmers and consumers.
  - o CSA Programs: Community Supported Agriculture (CSA) programs enable residents to subscribe to regular deliveries of fresh produce from local farms.
- Economic Development:
  - o Job Creation: Urban farming has created jobs and entrepreneurial opportunities, particularly for marginalized populations.
  - o Local Economy Boost: Money spent on locally grown produce stays within the community, enhancing the local economy.

- Environmental Benefits:
  - o Sustainability: Practices such as composting, rainwater harvesting, and organic farming contribute to environmental sustainability.
  - o Carbon Sequestration: Urban green spaces act as carbon sinks, playing a role in climate change mitigation.
- Policy and Support:

- o Urban Agriculture Ordinance: Detroit's Urban Agriculture Ordinance has provided a legal framework, making it easier to establish urban farms.
- o Grants and Support: Various grants, technical assistance, and resources are available to aspiring urban farmers.
- Challenges and Solutions:
  - o Land Tenure: Securing long-term access to land has been a challenge. Solutions include leasing arrangements and community land trusts.
  - o Water Access: Ensuring affordable water access for irrigation is an ongoing concern, addressed through innovative water-saving techniques.

Detroit's urban agriculture movement illustrates how cities can leverage underutilized resources to address complex challenges. The comprehensive approach, integrating social, economic, and environmental aspects, has turned Detroit into a model for other cities facing similar issues. The grassroots-driven success, supported by favorable policies, showcases urban agriculture's potential as a versatile tool for urban regeneration. Detroit's story serves as an inspiring case study, demonstrating how urban agriculture can be a catalyst for transformative change, fostering resilience, sustainability, and community well-being.

Singapore: A Model for Urban Agriculture Innovation

With limited land and a growing population, Singapore has embarked on a mission to enhance its food security and sustainability through urban agriculture. The city-state's unique challenges have spurred a wave of innovation, making Singapore a world leader in vertical farming, hydroponics, and other cutting-edge agricultural practices:

- Vertical Farming:

- o Sky Greens: The world's first low carbon, hydraulic-driven vertical farm, producing vegetables with minimal land, water, and energy.
  - o High-Density Production: Vertical farms allow for year-round cultivation in a small footprint, maximizing land use.
  - o Quality Control: Controlled environments ensure consistent, high-quality produce without the need for pesticides.
- Government Initiatives:
  - o "30 by 30" Goal: Singapore's government has set an ambitious goal to produce 30% of its nutritional needs locally by 2030.
  - o Grants and Funding: Generous grants are available to support local urban farming enterprises.
  - o Legislation and Policy Support: Favorable regulations and policy support have catalyzed growth in the urban agriculture sector.
- Aquaponics, Hydroponics, and Aeroponics:
  - o Innovation in Water-Based Growing: Singapore has embraced various water-based farming methods, allowing for soil-less cultivation in urban areas.
  - o Water and Resource Efficiency: These methods utilize up to 90% less water compared to conventional farming, aligning with sustainability goals.
- Community Involvement:
  - o Educational Programs: Numerous schools, institutions, and farms offer educational tours and workshops to teach urban farming skills.
  - o Community Gardens: The "Community in Bloom" initiative promotes gardening in public and private spaces, fostering community bonds.
- Technology and Research:

- o Smart Farming: Automation, sensors, and data analytics are used to optimize growing conditions and reduce labor costs.
- o Research Collaboration: Collaboration between universities, research institutions, and private enterprises drives continuous innovation in urban agriculture.

- Rooftop and Building Farming:
  - o Urban Redevelopment: Buildings and rooftops are being transformed into farming spaces, including commercial farms in office buildings.
  - o Green Roof Policy: Policies and incentives encourage building owners to establish green roofs, enhancing biodiversity and cooling the urban environment.
- Food Security and Sustainability:
  - o Reduced Dependence on Imports: Local production reduces dependence on food imports, enhancing resilience against global supply chain disruptions.
  - o Sustainable Practices: Emphasis on resource efficiency, waste reduction, and renewable energy aligns urban farming with broader sustainability goals.
- Challenges and Opportunities:
  - o Limited Space: The scarcity of land requires creative solutions and continued innovation.
  - o Energy Costs: High energy costs for controlled-environment agriculture are being addressed through renewable energy solutions and energy-efficient technologies.

Singapore's urban agriculture initiatives represent a remarkable example of how innovation, government support, community engagement, and technological advancement can converge to create a sustainable food system. The city-state's approach demonstrates the potential of urban agriculture in densely populated environments, even where land is scarce. Singapore's experience provides valuable insights and lessons for other cities seeking to integrate urban agriculture into their sustainability and food security strategies. The seamless blend of technology, policy, community, and environmental stewardship has positioned Singapore as a global frontrunner in urban agriculture, offering a replicable model for urban resilience and sustainable development.

Berlin, Germany: Community-Driven Urban Agriculture

Berlin's urban agriculture movement is a testament to community engagement, innovation, and a commitment to sustainability. With a rich history of community gardens and recent initiatives to support urban farming, Berlin has become a hub for urban agriculture enthusiasts and professionals alike:

- Community Gardens and Urban Farms:
    - Prinzessinnengärten: A mobile urban garden that educates the public about organic food production and urban living.
    - Tempelhofer Feld: Former airport turned public park, hosting community gardens and experimental agriculture projects.
    - Himmelbeet Community Garden: Offers gardening plots, workshops, and cultural events.
- Policy and Government Support:
    - Urban Agriculture Strategy: Berlin's government is developing strategies to promote urban agriculture, recognizing its value for the city.
    - Land Allocation: Various projects receive land for community gardens and urban farming activities.
- Education and Skill Development:

- o Urban Farmers Academy: Provides training and workshops for aspiring urban farmers.
  - o School Gardens: Several schools have integrated gardens into their curriculum.
- Economic Opportunities:
  - o Local Markets: Urban farms and community gardens contribute to local farmers' markets, supporting the local economy.
  - o Entrepreneurship: Initiatives like ECF Farmsystems support aquaponics startups and sustainable food production.
- Environmental Sustainability:
  - o Recycling and Composting: Many gardens and farms emphasize waste reduction, recycling, and composting.
  - o Biodiversity: Urban agriculture sites often contribute to biodiversity by providing habitats for various species.
- Social Cohesion and Well-Being:
  - o Community Building: Community gardens serve as social hubs, enhancing community cohesion and offering spaces for cultural events.
  - o Mental Health Benefits: Engagement with gardens and green spaces has shown positive effects on mental well-being.

- Technological Innovations:
  - o Rooftop Farms: Projects like ECF Farm Berlin explore rooftop farming and aquaponics.
  - o Smart Agriculture: Some farms employ smart technologies for efficient water and nutrient management.
- Challenges and Adaptations:
  - o Land Security: Land tenure and access remain a challenge for some community gardens.

- o Climate Considerations: Adapting to the specific challenges of Berlin's climate, such as cold winters, requires innovative approaches.
- Collaboration and Networks:
  - o Urban Agriculture Network Berlin: Coordinates projects, shares knowledge, and advocates for urban agriculture.
  - o International Collaboration: Berlin engages with other cities, sharing best practices in urban agriculture.

Berlin's approach to urban agriculture is rooted in community collaboration, innovation, and a genuine passion for sustainability. It reflects the collective desire of its citizens to engage with their food system, support local economies, enhance biodiversity, and foster social connections. From rooftop farming to school gardens, Berlin's diverse urban agriculture landscape is an inspiring example of how a city can embrace agriculture not only as a means to produce food but as a multifaceted tool for urban development. Berlin's urban agriculture model is a valuable case study for other cities aiming to cultivate a more sustainable, community-oriented urban environment.

Best Practices

From the case studies, a series of best practices have been identified to inspire other cities in developing urban agricultural initiatives.

Community Revitalization and Empowerment

Urban agriculture can rejuvenate communities by turning underused or abandoned spaces into thriving urban farms. By fostering community involvement, cities can create hubs of interaction that encourage social engagement, skills development, and a shared sense of purpose:

- Shared Spaces for Growth: Urban farms can become community centers where neighbors learn, grow, and share together.
- Community Empowerment: Empowering local communities through employment opportunities, entrepreneurship, and leadership in urban farming can lead to economic development and improved social cohesion.

Technological Advancements and Innovation

Innovation and technology in urban agriculture enable high-density farming in confined urban spaces. It maximizes efficiency while minimizing resource usage, creating a model for sustainable urban food production:

- Vertical Farming: Vertical systems allow for the productive use of vertical space, multiplying the yield per square foot.
- Rooftop Farms: Using rooftops for farming leverages otherwise wasted space, providing additional green coverage to cities.
- Water-Driven Systems: Advanced systems can utilize water efficiently, minimizing waste, and integrating with existing urban water management.

Government Support and Policy Integration

Government engagement is vital in establishing a robust urban farming sector. Policies and regulations can support urban agriculture, offering a structured and encouraging environment:

- Strategic Planning: Setting clear and ambitious goals can drive the sector forward, fostering innovation and alignment with citywide sustainability goals.
- Investment in Local Initiatives: Financial and logistical support for urban farming initiatives can enhance the growth and sustainability of these projects.

- Regulatory Framework: A well-defined regulatory framework can provide clarity and direction, encouraging compliance and innovation.

Education and Cultural Shifts

Education is a powerful tool for transforming urban farming into a central part of urban life. Cultivating a culture of sustainability through education can build a community of informed citizens engaged in urban agriculture:

- Workshops and Training: Offering workshops and hands-on training encourages participation and enhances skills.
- Youth Engagement: Engaging young people in urban farming fosters an understanding of food production, nutrition, and environmental stewardship from an early age.

Waste Management and Circular Economy Practices

A sustainable approach to waste can turn urban farming into an integral part of the city's circular economy, where waste is minimized and resources are reused:

- Composting and Recycling: Turning food waste into compost and utilizing recycled materials in urban farming supports sustainability.
- Biodiversity Support: Emphasizing diverse ecosystems and beneficial insects can reduce the need for pesticides and create a balanced urban ecosystem.

Community-Driven Innovation

Emphasizing community-driven initiatives can foster a sense of ownership and creativity within urban farming projects. It creates a grassroots movement that supports sustainable urban development:

- Community Gardens: Community gardens provide shared spaces for learning, socializing, and growing, fostering community bonds.
- Social Hubs: Social hubs focused on urban farming can become centers of education, collaboration, and innovation, leading to broader sustainability.

Health and Well-being

Urban farming can also play a crucial role in promoting health and well-being. The availability of fresh produce and opportunities for outdoor activities enhances physical health, while community engagement supports mental well-being:

- Nutrition and Food Security: Urban farms increase access to fresh and nutritious food, alleviating food deserts in urban areas.
- Mental and Physical Health Benefits: Participating in urban farming offers therapeutic benefits and encourages healthy outdoor activities.

Conclusion: A Holistic Approach to Urban Agriculture

The diverse best practices identified demonstrate a holistic approach to urban agriculture. By engaging communities, leveraging technology, aligning with government support, emphasizing education, embracing waste management practices, and fostering community-driven innovation, cities can create a vibrant and sustainable urban farming landscape.

These practices provide a roadmap adaptable to various urban contexts, reflecting the multifaceted value of urban agriculture. From economic development and social cohesion to environmental sustainability and health promotion, urban agriculture offers an integrative solution to many urban challenges.

For cities seeking to leverage the transformative potential of urban farming, these best practices offer a practical guide. By adopting and adapting these principles to local needs and resources, urban agriculture can become a central element of urban planning, shaping more resilient, sustainable, and engaged urban communities.

Chapter 5. Conclusion

Urban agriculture, as a vital component of green infrastructure, has emerged as a transformative practice that reshapes urban landscapes, revitalizes communities, and fosters sustainability. By integrating agriculture into the urban fabric, cities around the world have begun to harness the multifaceted benefits that urban agriculture offers, both to the environment and the population. This conclusion reflects on the overarching principles and practices that underline the significance of urban agriculture within green infrastructure.

First and foremost, urban agriculture has proven to be a powerful tool for community empowerment and social cohesion. From small community gardens to extensive rooftop farms, the urban agriculture movement has created spaces for community interaction, skill-building, and social engagement. These spaces not only promote local food production but also serve as hubs for education, entrepreneurship, and cultural expression.

Technological advancements have further accelerated the growth of urban agriculture, allowing for innovative practices such as vertical farming and advanced water management systems. These technologies have opened new possibilities for high-density, resource-efficient farming within the constraints of urban environments. The case studies of Detroit, Singapore, and Berlin have demonstrated the potential for cutting-edge technology to transform urban spaces into productive landscapes, aligning with broader sustainability goals.

Government support and policy integration have been key factors in the successful implementation of urban agriculture initiatives. A well-defined regulatory framework and strategic investment in local projects have provided the necessary foundation for growth and innovation.

Collaborative efforts between government agencies, private entities, community organizations, and educational institutions have fostered an environment conducive to the success of urban agriculture.

Education and a cultural shift towards sustainability have also played a crucial role in embedding urban agriculture within the fabric of city life. From school programs to community workshops, educational initiatives have raised awareness and cultivated a generation of urban farmers. These efforts have helped create a shift in perception, recognizing urban agriculture as an essential part of urban living and an effective means of addressing food security, health, and environmental challenges.

Waste management and circular economy practices have turned urban agriculture into a symbol of sustainable living. Through composting, recycling, and the implementation of diverse ecosystems, urban agriculture has contributed to the reduction of waste and the creation of self-sustaining, resilient systems. It has become a model for other sectors, demonstrating how circular principles can be integrated into urban life.

Furthermore, the exploration of health and well-being, land access and ownership, regulations and incentives, and sustainability goals has painted a comprehensive picture of the complex interplay between urban agriculture and broader city planning objectives. The emphasis on community-driven innovation, coupled with the recognition of challenges and the identification of solutions, has provided valuable insights for cities seeking to scale up urban agriculture.

In conclusion, urban agriculture, as part of green infrastructure, offers a multifaceted and integrative solution to urban challenges. From economic development to environmental stewardship, it creates synergies that enrich urban life and foster sustainable growth. This book has sought to provide a detailed examination of the principles, practices, and potentials of urban agriculture, emphasizing its role as a vital component of green infrastructure. The lessons learned from various case studies and the identification of best practices provide a roadmap for cities around the world, inspiring them to leverage urban agriculture as a key element of urban planning, shaping more resilient, inclusive, and sustainable urban futures.

Implications for Policy and Practice

Urban agriculture, as explored throughout this comprehensive study, presents not just an opportunity but an imperative for reshaping policy and practice. Integrating urban agriculture within the broader framework of green infrastructure has far-reaching implications that can profoundly impact the future of urban development. This concluding section on the implications for policy and practice distills key insights and proposes directions for cities worldwide to align their strategies with the principles of sustainability, resilience, and community engagement:

- Policy Integration and Collaboration: The successful implementation of urban agriculture necessitates a collaborative approach that transcends traditional silos within government. Creating a coherent policy framework that aligns with various sectors such as environment, health, transportation, and housing is vital. Cities must foster cross-departmental collaboration, ensuring that urban agriculture is not isolated but is a central aspect of urban planning and development.
- Regulatory Support and Incentives: To encourage urban agriculture initiatives, governments must provide clear regulatory guidelines, incentives, and support. This includes zoning laws that allow for urban farming, grants, and tax

incentives to support local projects, and streamlined permitting processes to enable innovation and expansion.

- Community Engagement and Empowerment: Urban agriculture is inherently a community-driven effort. Policies must reflect this by promoting inclusivity, empowering local communities, and recognizing the social value of urban agriculture. Partnerships with community organizations, local schools, and businesses can drive forward grassroots initiatives that resonate with local needs and aspirations.

- Education and Training: Building a culture of urban agriculture requires education at all levels. This encompasses school programs, community workshops, professional training, and public awareness campaigns. Ensuring that urban agriculture is embedded within the educational system fosters a generation of urban farmers and informed citizens who can champion sustainable practices.

- Technology and Innovation: Embracing technological advancements in urban agriculture, such as vertical farming, hydroponics, and intelligent water management, requires policies that support research, development, and the commercialization of new technologies. Governments and private sector collaboration can spur innovation, making urban agriculture more accessible, efficient, and scalable.

- Health and Well-being: Urban agriculture contributes to public health and well-being by increasing access to fresh, local produce, and providing recreational and therapeutic spaces. Policies should recognize these benefits, integrating urban agriculture into healthcare strategies, and promoting spaces for community wellness.

- Resilience and Climate Adaptation: As cities grapple with climate change, urban agriculture offers solutions for resilience and adaptation. Integrating urban farming within city-wide climate strategies ensures that urban landscapes are multifunctional, adaptable, and resilient to changing environmental conditions.

- Economic Development: Urban agriculture can be a catalyst for local economic growth, providing employment opportunities, stimulating local markets, and revitalizing underutilized spaces. Policy must recognize and leverage these economic potentials, positioning urban agriculture as a key driver of sustainable economic development.

- Monitoring and Evaluation: Ongoing monitoring and evaluation are crucial to understand the impacts of urban agriculture and to refine policies over time. Implementing robust data collection and assessment mechanisms provides insights into successes, challenges, and areas for improvement.

- Global Knowledge Exchange: Urban agriculture is a global movement, and cities can learn from each other's experiences. Encouraging international collaboration, knowledge sharing, and benchmarking can inspire and guide local policy and practice.

In summary, the implications for policy and practice in urban agriculture are multifaceted and profound. This book's exploration has laid the groundwork for understanding how urban agriculture can be effectively integrated into urban policy, planning, and practice. By addressing the complexities and recognizing the potential, cities can leverage urban agriculture as a versatile tool to advance sustainability, resilience, equity, and community well-being. The road ahead is filled with opportunity and challenge, and the insights gleaned from this comprehensive study provide a valuable guide for policymakers, practitioners, and citizens alike as they navigate the evolving landscape of urban agriculture within green infrastructure.

Vision for the Future

The vision for the future of urban agriculture within green infrastructure is a vibrant, transformative, and holistic one, interweaving ecology, community, economy, and culture into the urban fabric. It represents not just an agricultural shift but a societal evolution, redefining how cities grow, thrive, and sustain themselves. In this conclusion, we will sketch a vision that encapsulates the key principles, opportunities, and hopes that urban agriculture presents for our collective urban future.

A Resilient Ecosystem

The future vision of urban agriculture encompasses a resilient urban ecosystem where food production is a core function of the city's landscape. Integrated within green infrastructure, urban agriculture becomes a linchpin for environmental sustainability, water management, biodiversity, and climate adaptation. Rooftop farms, vertical gardens, community plots, and urban orchards become common features of the urban environment, not just beautifying the city but enhancing its ecological resilience.

Community Empowerment and Well-being

Urban agriculture fosters a sense of community ownership, engagement, and well-being. It creates spaces for social interaction, learning, and therapy, bridging divides and cultivating a sense of shared purpose. The future city is a place where every citizen has access to fresh, healthy food, and opportunities to participate in growing it, regardless of socio-economic status.

Educational Revolution

Education plays a central role in this vision, with schools, colleges, and community centers becoming hubs for agricultural innovation, knowledge sharing, and skill development. Urban agriculture is taught, experienced, and celebrated, nurturing a new generation of urban farmers, ecologists, and conscious consumers.

Economic Opportunities and Circular Economy

Urban agriculture in the future city creates diverse economic opportunities, from small-scale community gardens to commercial urban farms. It drives a circular economy where waste is minimized, and resources are recycled. It stimulates local economies, generating employment, revitalizing neighborhoods, and creating markets for local produce.

Technological Innovation

The future of urban agriculture is underpinned by technological innovation that enhances productivity, sustainability, and accessibility. Smart farming technologies, renewable energy, water-saving practices, and scientific advancements in soil and plant health are integral to this future, guided by principles of sustainability.

Policy Alignment and Global Leadership

At the governance level, this vision requires alignment of policies across various sectors, supportive regulations, and public investments. Cities become leaders in sustainable development, setting examples for others to follow, and influencing national and international policy in favor of urban agriculture.

Cultural Shift and Aesthetic Transformation

Urban agriculture in this vision is not just functional but aesthetic and cultural. It shapes the urban identity, adds character to the cityscape, and fosters a cultural connection to the land and food. Cities become living landscapes that celebrate growth, harvest, and seasonality.

Health and Quality of Life

The integration of urban agriculture enhances public health through improved nutrition, mental well-being, and physical activity. It creates spaces for recreation, relaxation, and connection to nature within the bustling urban environment.

Global Collaboration and Knowledge Exchange

The future of urban agriculture is globally connected, with cities collaborating, sharing knowledge, and learning from each other's successes and challenges. It's a global movement towards a more sustainable, humane, and nourishing urban life.

In conclusion, the vision for the future of urban agriculture within green infrastructure is a compelling and optimistic one. It paints a picture of cities that are ecologically resilient, socially just, economically vibrant, and deeply connected to the natural world. It challenges us to rethink how we design, govern, and live in our urban environments. It invites us to embrace urban agriculture not as a trend but as a transformative path towards a more sustainable, inclusive, and humane urban future. This vision is not merely a dream but a tangible goal that can be achieved through concerted effort, innovation, leadership, and above all, a shared belief in the power of urban agriculture to reshape our cities and our lives.

Closing Thoughts

The journey through urban agriculture and its integration within green infrastructure has unveiled a multifaceted and transformative approach to city living. As we stand at the crossroads of unprecedented urbanization and environmental challenges, urban agriculture emerges as more than a solution; it's a vision for a new way of life. It embodies a future where cities are not only places of dwelling but thriving ecosystems that nourish the body, soul, and planet. The lessons, case studies, and insights shared in this book serve as both inspiration and a call to action. It is an invitation to policymakers, practitioners, communities, and individuals to embrace this vision, work collaboratively, innovate, and pave the way for a future where urban agriculture is not an exception but a defining feature of our urban landscape. The seeds have been sown; now, it's time to cultivate our shared urban garden.